MYSTERIES
OF THE
UNIVERSE

Answerable and Unanswerable Questions

Peter Altman

This UK edition is published and distributed non-exclusively by:
Altman Publishing
7 Ash Copse, Bricket Wood, St Albans AL2 3YA
altmanpublishing@virginmedia.com

© Peter Altman

Original edition published by:
Jaico Publishing House
A-2 Jash Chambers, 7-A Sir Phirozshah Mehta Road
Fort, Mumbai - 400 001
jaicopub@jaicobooks.com
www.jaicobooks.com

All rights reserved by Jaico Publishing House.

MYSTERIES OF THE UNIVERSE
ISBN 978-1-86036-062-6

First Impression: 2021

No part of this book may be reproduced or utilized in
any form or by any means, electronic or
mechanical including photocopying, recording or by any
information storage and retrieval system, without
permission in writing from the Jaico Publishing House.

Page design and layout: R. Ajith Kumar, Delhi, India

For Joan
(but you don't have to read it)

CONTENTS

Author's note ... ix
Preface ... xi
Introduction: Answerable and unanswerable questions ... xiii

THE QUESTIONS

Q1 How did the universe begin? ... 3
Q2 Do other universes exist? (and what is the weight of an orange?) ... 31
Q3 How did life begin? ... 38
Q4 How old are the Earth and the universe? ... 57
Q5 Does alien life exist? ... 80
Q6 Could we travel to other stars and galaxies? ... 161
Q7 Is time travel possible? ... 173
Q8 Could we live longer? ... 181
Q9 How and when will the world end? ... 192
Q10 Is the future predetermined? ... 220
Q11 Does prayer work? ... 234
Q12 Astrology: Sense or nonsense? ... 241
Q13 Will we ever know everything? ... 248
Q14 The Creator question ... 258

Appendix

A.	*Evidence and how to assess it*	265
B.	*The cosmic distance ladder—how astronomers measure distance*	271

Final comments — 292
Free Gift – The Inescapable Conclusion — 293
Index — 300

AUTHOR'S NOTE

The self-imposed rule that governs the content of this book is that answers are based, as far as possible, on scientific evidence. The opinions are entirely my own and although I have tried to base them on established facts and principles, some people may disagree with my conclusions. That is fine, as it leads to interesting debates and discussions.

Chapter titles in this book are in the form of questions which are addressed in the text that follows. The chapters end either with the conclusion that 'this is an answerable question' or that 'this is an unanswerable question'.

I consider a question to be answerable if there is sufficient evidence to provide an answer. I consider a question to be unanswerable if there is insufficient evidence to provide an answer. In these cases, I have provided what I call a 'best guess answer'.

I have attempted to contact copyright holders where illustrations are either not my own or where I believe them not to be in the public domain, and I would like to thank all those who have kindly given permission for the use of their material. In a few cases, I have been unsuccessful in contacting the concerned people/organizations and although attribution has been given in the legends, I apologize if I have unwittingly infringed any copyrights; such oversights will be corrected in any future editions if I am made aware of them.

To save confusion, since different naming conventions exist in different countries, the following have been used in this book:

1,000,000,000	1 billion	10^9
1,000,000,000,000	1 trillion	10^{12}
1,000,000,000,000,000	1 quadrillion	10^{15}

Since each chapter deals with a different topic, the book does not have to be read in sequence—there is no reason why you cannot start with the question that interests you the most and go back to other questions later. Appendix B is rather mathematical and is only included for those who may be interested in how astronomical distances are measured.

One more thing—some parts of this book relate to matters of religion. It is absolutely not my intention to disrespect anyone's beliefs and in this regard the comments I have made are entirely based on my own opinions. I hope that they will be read in that context.

I would also like to thank Jaico Publishing House for having faith in me and agreeing to publish my book. It has been a pleasure to work with their team. And finally, as a token of appreciation for buying the book, I am pleased to offer you a free gift on page 293. I hope you enjoy it.

PREFACE

The universe is a really big place. Current estimates say it is 92 billion light years in diameter, and getting bigger. Have you ever wondered how it all started and how it might end, or whether there is alien life on other planets, or whether time travel is possible and how life began? There must be answers to these and other fascinating questions about life and the universe, but in most cases, we do not know what these answers are. However, that does not mean that we cannot make a best guess based on what we do know.

That is what this book is about. It is a journey that allows us to wander through the universe and wonder about the answers to many questions. Although a few of the questions can be answered with a fair degree of certainty, most are unanswerable. All we can do therefore is give best guess answers based on current knowledge.

INTRODUCTION

ANSWERABLE AND UNANSWERABLE QUESTIONS

Questions are often categorized according to their degree of difficulty. Here are two that most people would call easy:

> What is the capital of India?
> New Delhi
> What is the square root of 100?
> 10

Conversely, here are two questions that most people would call difficult:

> What is the largest moon of Pluto?
> Charon
> What was Rembrandt's full name?
> Rembrandt Harmenszoon van Rijn

The distinction between easy and difficult is, of course, subjective. Any question to which you know the answer would, for you, be easy, while one to which you do not would be difficult. However, the four sample questions here all have one thing in common which is that they all have a definitive answer. You may know the answer but if you don't, you can look it up because an answer does exist. They are all answerable questions.

Now compare this with these questions:

Who was Jack the Ripper?
When will Prince George become King of the United Kingdom (UK)?

There are no definitive answers to these questions. One might be able to give an educated guess based on some research but that is about the best one could do. The first question deals with a past event and although there must be an answer, the information is no longer available so no correct answer can now be given. The second question deals with a future event so there can be no answer, just a best guess.

For the purposes of this book and the questions that it poses, we shall categorize the questions as follows:

An answerable question is one which, based on available information, is answerable with a reasonable degree of certainty.

An unanswerable question is one which, even with the benefit of available information, is still unanswerable with a reasonable degree of certainty. In these cases, all one can do is make a best guess based on whatever information happens to be available.

Some answers are easier to arrive at than others, depending on the available evidence. Good evidence gives us confidence in our answer and poor or little evidence makes the answer more of a guess. In some cases, that is the best we can do.

THE QUESTIONS

Image 2. *The Thinker (Rodin, c. 1902).*

Q1

HOW DID THE UNIVERSE BEGIN?

We live on a planet (Earth) which circles a star (the Sun) which is part of a galaxy (the Milky Way). A drawing of what the Milky Way might look like if viewed from above and outside the galaxy is here.

Image 3. *Drawing of the Milky Way as seen from above and outside the galaxy. The position of the Sun is shown by a white dot. ©NASA.*

Our Sun is located on one of the spiral arms (the Orion Arm) and is about 25,000 light years from the centre of the galaxy.

Galaxies are huge objects. Ours is about 100,000 light years wide (that means it would take light 100,000 years to go from one side to the other, equivalent to about 600,000 million million miles) and about 1,000 light years thick, and contains about 500 billion stars.

When you look up at the sky on a clear night, away from bright city lights, you might see a band of hazy light stretching across the sky as shown in the photograph below. This is one of the spiral arms of the Milky Way galaxy.

Some galaxies have neighbours (in the astronomical sense, since they can be millions of light years away) and such a group of galaxies is called a cluster. Some clusters can even form superclusters. Some galaxies are spiral, like ours, and others can be elliptical or other shapes.

Current estimates put the total number of galaxies in the universe at about 100 billion. The space between galaxies—

Image 4. *The Milky Way as seen on a clear night. ©NASA.*

the intergalactic space—is pretty empty with only a few atoms drifting about and also very cold at a temperature of about -270°C. The oldest stars and galaxies are over 13 billion years old as measured by a variety of astronomical techniques.

SIZE MATTERS

Our universe is a really big place, so big in fact that its size is impossible to contemplate. It is easy enough to visualize a distance of a few miles or kilometres. You may drive 10 miles to visit a restaurant, or perhaps 100 miles to go to a nearby city. Even distances of a few thousand miles are fairly easy to comprehend during air travel to other continents.

But what about distances of millions of miles? These are so far removed from our usual activities that we have no personal experience by which to gauge them. In cosmological terms however, a few million miles is no distance at all. It will just about take you to the nearest planet (Venus, which is about 25 million miles away at its closest approach). You would need to travel over 3 billion (3,000,000,000) miles to reach the outposts of our Solar System and a staggering 25 trillion (25,000,000,000,000) miles to reach the nearest star (Proxima Centauri).

The numbers get even bigger when we consider the size of our galaxy, the Milky Way, which is about 600 quadrillion (600,000,000,000,000,000) miles wide.

Many analogies have been created to make such numbers meaningful. Here are two examples: if the Earth is represented by a marble, then the edge of the Solar System would be over 5 km (3 miles) away. Or, if you were travelling at jet airplane speed, it would take about 700 years to reach Pluto and over 5 million years to reach our nearest star Proxima Centauri.

Image 5. *"This is the Captain speaking. We shall be landing in 688 years."* ©*Author.*

But these devices do not help much since it is impossible to form a mental image of the reality of these vast distances.

So how big is the universe? Current evidence shows that the diameter of the observable universe is approximately 92 billion light years, and it is getting bigger.

What is meant by the term 'observable universe'? Simply put, it means the universe that we can see. There may or may not be parts that we cannot see. But why, if they are there, might we be unable to see them?

That is a fair question. It is known that the universe is expanding and that the galaxies are rushing away from each other. This is because space itself is expanding.

It turns out that this expansion of space can be very fast indeed, even exceeding the speed of light. This does not contradict Albert Einstein's theories since no material object is travelling at a super-luminal speed, only space which is nothing. See images 6 and 7 on the next page.

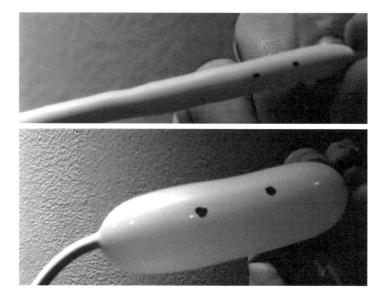

Images 6 & 7. *Think of the surface of a balloon as it is being filled with air. Two marks made on the surface of the balloon would start to grow further apart as the balloon is blown up since the surface of the balloon is expanding. In the top photograph, the balloon is only slightly inflated and the marks are about 20mm apart. In the bottom photograph where the balloon is more inflated, the marks have moved further apart and there is about a 50mm separation. The marks themselves have not moved but they are now further apart because the balloon material has expanded. This is a reasonable way in which to visualize the expansion of space.* ©*Author.*

Now we can get back to the question of the observable universe and why everything that exists may not be visible to us. Imagine that you are standing at the end of a slowly moving train and you throw a ball out to a dog standing on the track.

If the train is moving away from the dog at 16 kph (10 mph) and the speed of the ball, relative to the thrower, is 24 kph (15 mph), it will reach the dog at 8 kph (5 mph), assuming of course that it does not hit the ground first.

Image 8. *A dog on a railway track waiting for a ball. ©Author.*

Now imagine that the train is moving away at 32 kph (20 mph) but the speed of the ball is still 24 kph (15 mph). Under these conditions the ball will never reach him. Obviously, he could walk along the track to pick it up but that is not the point. Staying where he is, he will never get the ball because it will always be moving away from him at 8 kph (5 mph) until it hits the ground.

It is the same with galaxies that are moving away from us at speeds exceeding that of light due to the expansion of the space between the galaxies and us. The light from those galaxies will never reach us so we can never see them. We have no way of knowing whether such galaxies or other objects exist or not, and that is why we call the universe we can see the 'observable' universe. Whether or not anything exists beyond that we cannot say.

This raises the interesting question of what might exist beyond this 'event horizon' which we cannot breach. Obviously, we can only speculate. On a simplistic level, there would seem to be two possibilities: more universe, or nothing. If there is more universe, then what happens when we reach a position where there is no more universe? Some might say that this would never happen, arguing that the universe continues forever, and is infinite in size. That is a difficult concept to grasp but that is perhaps because we are just not able to comprehend what this means.

On the other hand, maybe there is a point where there is nothing at all no matter how much further you look. Existing galaxies may be hurtling towards this empty space but then there will be further empty space beyond. So, does the empty space continue forever? Some might say that space, even when it is empty, is still part of our universe. What lies beyond empty space? Perhaps one day there will be a way to find out but for now that really is an unanswerable question.

Let us now return to the known part of the universe before we consider how it might have all started. Objects such as the Andromeda nebula are just about visible to the naked eye and have been known since antiquity. Many other nebulae had been discovered with the aid of the telescope, and up to the 1920s, it was thought that all these objects were located within our own Milky Way galaxy.

In 1924, Edwin Hubble, an American astronomer, used a newly-discovered method for measuring stellar distances and calculated that the Andromeda nebula was about 1 million light years away (the currently accepted value is about 2.5 million but he had made his point). Since the diameter of the Milky Way galaxy is about 100,000 light years, Andromeda must lie outside our galaxy (more on this in Appendix B).

Hubble had therefore proved that Andromeda and all the other nebulae lay outside our own galaxy and that they were separate galaxies in their own right.

Image 9. *Edwin Hubble, 1889–1953. ©NASA.*

Hubble also showed that these other galaxies were receding from us in all directions, giving rise to the conclusion that the universe was expanding. The further the galaxies were from us, the faster they were receding.

This made it possible to calculate backwards to the time when the universe started to expand, and this figure would represent the age of the universe. Current estimates give this age as close to 13.7 billion years.

This implies that about 13.7 billion years ago the universe consisted of a single point (known as a 'singularity') that, for some reason, became unstable and exploded into all the matter that we see today.

What this single point was, where it was, how and why it exploded, are questions we cannot answer except to say that given a certain set of circumstances it had to happen. No one made it happen, it just did because of the laws of physics. It may have been very unlikely but that does not matter since once was enough.

This is known as the Big Bang theory of the origin of the universe. Although there are a number of theories about the mechanism of creating something from nothing, none of them provide an answer that is satisfactory in layman's terms as to how this could have come about.

Visualizing the Big Bang is not easy. It is tempting to think of a massive explosion that scattered all of the matter in the universe into the surrounding empty space but, according to current thinking, this is not the way to look at it.

Cosmologists will say that we should imagine a vast number of Big Bangs occurring simultaneously in many different places. We should also imagine, although this is really difficult to do, that space and time came into being at the moment of the Big Bang. As Brian Cox puts it in his book *Wonders of the Universe* co-written with Andrew Cohen, "Everywhere that is here now was there then but just squashed up". But if space is squashed up, then what is outside it? That is a question for a cosmologist but I suspect that only another cosmologist would comprehend the answer.

Let us consider the evidence for the Big Bang theory. A theory is an attempt to explain something that has happened which can then be tested by experiment.

For example, if a light suddenly goes out, we might expect to find a broken filament if we examine the bulb. In the same way, many scientific theories enable predictions to be made which can then be tested. If the predictions turn out to be true, then this lends support to the correctness of the theory. Experimental support for the Big Bang theory came in 1992 as a result of NASA's Cosmic Background Explorer (COBE) satellite.

If you look at the night sky with an optical telescope, the space between heavenly objects—planets, stars and galaxies—is black. However, if you look with a radio telescope, then you can detect a faint glow. This glow is not due to any planets, stars or galaxies and is, therefore, called 'background radiation'. It is strongest in the microwave region of the electromagnetic spectrum. Because of this, it is known as the Cosmic Microwave Background (CMB) radiation. The existence of CMB was predicted in 1948 and discovered mostly by accident in 1964 by two American radio astronomers, Arno Penzias and Robert Wilson. It earned them the Nobel Prize for Physics in 1978.

When NASA's COBE satellite made very detailed measurements of the CMB, it found small temperature variations in what was originally thought to be a completely uniform glow. These variations were visualized as a temperature map of the sky and published in 1992. The variations were small—of the order of a few parts in 100,000—but they were there.

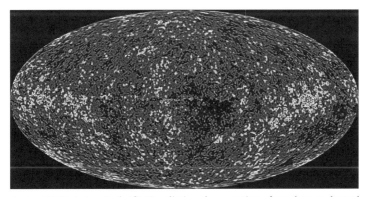

Image 10. Variations in the CMB radiation, shown as tints of grey here, as detected by the COBE satellite. ©NASA, 1992.

The map shown in Image 10 was published in newspapers and magazines around the world (in full colour but shown here in shades of black and white).

Why was this map so significant? The temperature variations, shown as different shades of grey here but in colour in the original map, are precisely what would be expected if the universe began as the result of a Big Bang. No other theory apart from the Big Bang theory has yet been proposed that can explain these variations.

Current knowledge therefore has led astrophysicists and cosmologists to propose that the universe did indeed begin its life as the result of a Big Bang, and that this was followed by a period of very rapid expansion known as inflation. This is depicted in Image 11.

The timescale along the bottom of the diagram indicates how rapid this expansion is assumed to have been, eventually slowing down to what is observed today.

Where do these timings come from? Detailed studies of the CMB, and also of nuclear reactions inside devices such as the Large Hadron Collider, have enabled physicists to estimate the time it takes for certain sub-atomic reactions

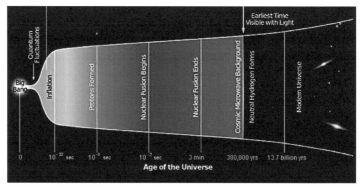

Image 11. Diagrammatic representation of cosmic inflation. ©Wikimedia Commons.

to take place. Such studies have allowed cosmologists to estimate the timescale of reactions in the early universe. What is the evidence in favour of inflation? This is a complex subject on which entire books have been written so we will mention just one piece of evidence, sometimes known as the 'cosmic conspiracy'.

On a broad scale, the universe looks much the same in all directions and is also at a similar temperature in all directions. If we consider a pair of galaxies, each in opposite directions as viewed from Earth and each 10 billion light years from Earth, then the galaxies would be 20 billion light years apart.

As stated above, each galaxy and its neighbourhood would look similar and be at a similar temperature. The universe is less than 20 billion years old, so there would not yet have been enough time for radiation or gravitational effects to pass from one region to the other (since travelling at light speed the 20 billion light years' distance of separation would take 20 billion years to traverse).

How then can we explain the similarity between these two regions given that there is insufficient time for temperature and other physical properties to have equalised themselves? One way is to speculate that the materials from which the two galaxies were created were initially in very close contact (and therefore equalised in temperature and other properties) and then expanded very rapidly.

Consider two cups of coffee, one in a coffee bar in London and the other in a coffee bar in New York. It would be surprising if both coffees were at precisely the same temperature and contained precisely the same amounts and types of coffee since they are clearly too far apart to influence each other.

One explanation, admittedly rather bizarre in an everyday

London

New York

context, would be that both coffees were initially from the same machine (and therefore were of the same type and at the same temperature) but they have, somehow and very quickly, become separated by 3,000 miles. This is a very simplified scenario that may help to put the inflation theory into an everyday context.

Having said that, there is increasing interest amongst cosmologists to provide an answer to the obvious next question—what happened before the Big Bang?

And if all the material in the universe came from Big Bang explosions, then where did the original material that exploded come from?

We can consider three options:
1 The universe arose on its own (Big Bang or other process)
2 The universe has always existed
3 The universe was the work of a Creator

1. THE UNIVERSE AROSE ON ITS OWN

If the universe arose on its own, how did this happen? We do not know, but there are some clues from things that we do know.

Let us imagine a situation just before the universe came into being.

There is nothing and then suddenly there is something. Why? Something happened, we do not know what, but it resulted in the appearance of all the material that we see in the universe today although not necessarily in the same form. Try and relate this to, for example, visible light being split into the seven colours of the spectrum after it passes through a prism. At first there is just white light and then suddenly there are seven colours because something happened.

Or, your car goes around a bend too fast and runs off the road. You are driving along and suddenly, because something happens (going too fast), you are off the road. These changes are due to the inherent properties of light and the car.

It is hard to imagine all of this, but it belongs to the realms of theoretical physics and higher mathematics so is always going to be difficult for the non-specialist to visualize.

Irrespective of the conditions, it was inevitable that the universe would come into being because of the inherent properties of whatever existed at that moment, even if in layman's terms, it was 'nothing'. We also need to remember Einstein's famous equation $E = mc^2$ (E = energy; m = mass; c = the speed of light) which tells us that mass and energy are interchangeable so that an amount of energy could have been turned into an amount of mass.

But then where did the energy come from? This is another good question with no satisfactory answer at present, even amongst cosmologists. But that does not mean it is not there—it just means that we do not know what it is.

2. THE UNIVERSE HAS ALWAYS EXISTED

Saying that the universe has always existed immediately removes the problem of how it started. It did not start, it has always existed.

This is still consistent with the Big Bang theory and just implies that all the matter that existed in the point that exploded at the Big Bang has always existed. The universe may be an oscillating system going from Big Bang to Big Crunch every trillion years or so, repeating the process forever (see Question 9). It has always existed and always will exist.

The concept of something having always existed is hard to visualise. We cannot ask 'Where did it come from?' because it did not come from anywhere – it has just always been there although not necessarily in the same form.

The concept of something having always existed is not an easy one to visualise or to accept, and it certainly will not be an easy one to test or prove.

3. THE UNIVERSE WAS THE WORK OF A CREATOR

If you think about it, what does it actually mean to say that the universe arose through the work of a creator or designer, or in other words, a God? Like the concept of life arising not on Earth but extra-terrestrially and brought here by a meteorite or comet or asteroid (see Question 3), it does not solve the problem but just places it somewhere else.

If the universe did result from the work of a Creator, how was it done? There would have to have been a mechanism. Saying that humans cannot or are not meant to understand it does not help and just pushes the question away.

So, whether the universe arose by itself as a result of the laws of physics, or has always existed, or arose as a result of the work of a Creator, we are still faced with the same question of how it was done. Invoking a creator or designer does not help with solving this problem. So why bother?

The universe exists. We do not know whether it came into being at the moment of the Big Bang or whether it has always existed. What we do know is that it is here now. We can at least all agree on that, whatever our faith or belief. Some of us say a Creator did it; some say it happened on its own due to the laws of physics because under a certain situation it had to happen; and some say that it has always existed. (See also Question 14).

Everything that has ever been studied obeys the fundamental laws of nature. Obviously, we cannot say that those laws were exactly the same 13.7 billion years ago at the moment of the Big Bang as they are now since no one was there to investigate them. But even if they were different, it seems reasonable to think that there were laws that governed how matter behaved.

UNDISCOVERED LAWS

What about as yet undiscovered laws? Could there be some laws of physics that we do not know about? Yes, of course there could.

However, these are likely to be at the very extremes of nature, such as at very low or very high temperatures or pressures, or involving very small or very large objects. These are all situations that are difficult to study. It is most unlikely that any as yet undiscovered laws of physics would impact everyday situations. Why? Because we would know about them already from our everyday experiences.

We have seen that things happen in predictable ways. If, for example, there was an unknown law about how centrifugal force works, then this could obviously not be taken into account when engineers work out how to construct a fairground roundabout. Riders would either not experience anything or would be thrown off. The fact that this does not happen means that for all practical purposes, we do know all about centrifugal force.

Image 13. *"Do you want another go, Grandma?"* © *Author.*

As is clear from Image 13, this fairground ride is doing its job properly and my granddaughter wants it to continue and my wife cannot wait for it to end.

Another example is the notion of a twin Earth planet on the other side of the Sun which is forever hidden from our view. This is an old idea dating back many thousands of years, one that even spawned the 1950s American science fiction comic *Twin Earths* (Image 14).

We knew that such a planet did not exist long before spacecraft confirmed it visually. How? Because if it was there, the calculations made by astronomers on the motions of the other planets would be wrong since any gravitational effects of the putative planet would not have been taken into account.

This example leads us into an interesting discussion and

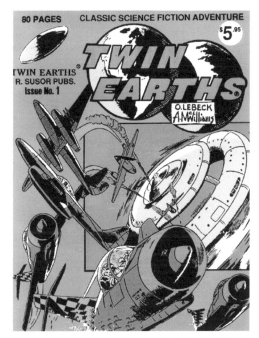

Image 14. Twin Earths comic No.1. ©R. Susor Publications.

shows how an undiscovered law of physics was eventually found.

It all started with the planet Mercury. Studies of the planet's orbit around the sun never quite matched the predictions based on calculations of the orbit. In 1840, French mathematician and astronomer Urbain Le Verrier was given the task of trying to find out why.

Nearly 20 years later, Le Verrier published an extremely thorough study of Mercury's orbit based on numerous meticulous observations.

In 1846, Le Verrier was involved in the discovery of the planet Neptune. Uranus, the seventh planet from the sun, was the first 'new' planet to be discovered, the other six having been known from antiquity. William Herschel, a

Image 15. *Urbain Le Verrier, 1811–77. ©Wikimedia Commons.*

British astronomer, discovered the planet on 13 March 1781 and received worldwide acclaim. Many astronomers then attempted to calculate the orbit of the new planet but it did not behave as they predicted.

Some people suggested that Isaac Newton's Laws of Motion did not apply so far away from the Sun (nearly 2 billion miles).

Others suggested that perhaps the orbit was being affected by the gravitational pull of yet another unknown planet even further away.

Calculations were made as to where such a planet should be in order to have such an effect, and on 23 September 1846, the planet Neptune was discovered almost exactly where it was predicted. Various astronomers share the credit for this—Le Verrier who did the calculations, and the German astronomers Johann Galle and Heinrich Louis d'Arrest who looked for and found the planet where Le Verrier predicted it would be.

Further studies of the Uranian and Neptunian orbits suggested that there might be a further unknown planet outside the orbit of Neptune. This eventually led to the discovery of Pluto by the American astronomer Clyde Tombaugh on 18 February 1930.

In view of his success with discovering Neptune, it is perhaps not surprising that Le Verrier suggested that the discrepancy in Mercury's orbit was due to the existence of an unknown planet between Mercury and the Sun. He proposed the name Vulcan for this planet, Vulcan being the Roman god of volcanoes and fire. This was thought to be an appropriate name for an object so close to the Sun.

The fictional planet Vulcan of *Star Trek* fame was supposedly in orbit around the star 40 Eridani, about 16 light years from Earth. As I was revising the manuscript, there was an announcement from the Royal Astronomical Society that a planet has been discovered circling 40 Eridani. Hullo, Mr. Spock!

Given Le Verrier's reputation, astronomers around the world started looking for this new planet but without success. Le Verrier died in 1877 convinced that he had discovered a second planet. However, since close to 20 years of observations by astronomers in many countries failed to find the new planet, the consensus was that it did not exist. But if Vulcan did not exist, how then could the discrepancies in Mercury's orbit be explained?

The answer came around 40 years later when, in 1915, Albert Einstein published his work on the Theory of Relativity.

Einstein provided a new approach to the understanding of gravity as compared to the Newtonian Laws of Motion.

The apparent discrepancies in the orbit of Mercury were fully explained by the application of Einsteinian rather than Newtonian mechanics and the need for an extra planet disappeared.

Relativity also resulted in changes to the calculated orbits of the other planets but since these were much further from the Sun than Mercury, the differences between the observed and calculated orbits were minor and hence within accepted margins of error. In 1915, therefore, a new law of physics was discovered which explained observations that previously were unexplainable.

MAIN POINT

It is most unlikely that there are unknown laws of physics still to be discovered that will have significant effects on how most things behave. It is only at the very extremes of nature, that is, the unbelievably small, large, hot or cold, that we are likely to discover something really new. Having said that however, it would be wrong to dismiss the concepts of Dark Matter and Dark Energy (see Question 9); an understanding of these currently theoretical entities may indeed lead to revised knowledge about the fundamental particles of nature. Who knows what properties they may have that we could exploit?

We can return now to our discussion of the origin of the universe. If the universe, as a finished article, obeys the laws of physics, then it is reasonable to assume that the universe, as a not-yet existing article but an about-to-appear article, also obeyed the laws of physics even if those laws were not the same as they are now.

In other words, as we have said before, under the right conditions, it had to arise. A Creator, if he/she was there at all, had nothing to do except to watch and admire.

So how did this actually work? If it arose rather than always having existed, how could something arise out of nothing?

Let us digress slightly before we try and address this question.

Ideas about the structure of matter have undergone many changes over time. The early understanding was that everything was made of just four elements—earth, air, fire and water.

Much later, in the 18th century, chemists began to realize that these were not elements at all, and started to isolate and list what we now recognize as elements, such as hydrogen, oxygen, phosphorus, sulphur, etc.

Then, at the end of the 19th century, it was found that elements could still be subdivided into even smaller particles, that is, the electrons, protons and neutrons that make up each atom.

The development of powerful particle accelerators in the 1950s led to the discovery of a huge number of even smaller particles which spawned the entirely new subject of particle physics. Quarks, mesons, baryons, hadrons and bosons are just some of the entities that have been discovered. And so, it goes on.

It is difficult if not impossible for a layman to form a mental image of what these latest particles 'look' like or are made of. It was easy enough for the atom—a sort of mini solar system with a central nucleus and electrons spinning around it. But how do you picture a boson?

Image 16 on the next page is about the best we can do. It is an image from a Spark Chamber, a device used to visualise what happens when particles are made to collide with each other at high speeds and high energy.

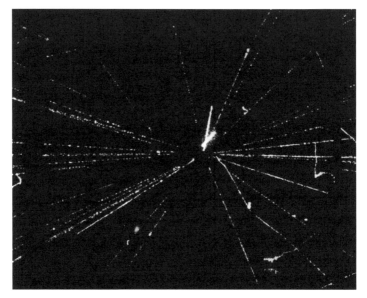

Image 16. *Spark chamber showing particle collisions.* ©*Wikimedia Commons.*

The collision is seen at the centre of the image and the tracks of the resulting smaller particles are seen to be radiating outwards. The angle and energy of these new particles can then be used to characterise them. So, although we cannot actually see the particles, we can at least see their tracks, so we know they exist. In addition to all these new particles, there are also four basic forces that need to be considered.

These are gravity, a long-range force that acts as an attraction on all matter; electromagnetic force, a long-range force that acts on particles that have an electric charge; strong nuclear force, the force that holds the protons and neutrons together inside the nucleus of an atom; and weak nuclear force, the force that causes atoms to disintegrate as in radioactivity.

Scientists from Einstein onwards have long been searching for a Theory of Everything—a Unified Theory that brings together all that is known about the many particles that make up matter and the four forces just described. It has become a Holy Grail of physics.

In the 1980s, a mathematical model called String Theory was developed. Briefly, mathematical equations were constructed that appeared to show how all the known particles and forces could be combined and described by one-dimensional strings—small units that had only length and no height or width.

To make it even more difficult to visualise, these strings existed in 11 dimensions! This became known as M Theory and it is currently thought by some scientists to represent the elusive Unified Theory of Everything. It also, apparently, allows for something to be created out of nothing although the explanation of how this could happen is firmly in the realms of theoretical physics and advanced mathematics, and would certainly be unconvincing to anyone not immersed in this field of research.

Some people might find this explanation unsatisfactory. To phrase the question at its most basic level, how could all of the vast amount of material in the universe have arisen from 'nothing'? Look at it this way. Just because we do not have an answer to a question does not mean that one is not there; it just means that we do not know what it is.

We can illustrate this point with the description of a card trick. The magician asks you to choose a card and then tells you what it is. How is it done? Here are some possibilities:

Mysteries of the Universe 27

> **All the cards in the pack are the same**

No, because you can examine the pack.

> **The magician forces the card on you**

No, because you hold the pack and take the card you want.

> **The magician glimpses the card as you take it**

No, because you take the card as you hold the pack under the table.

> **The cards are marked**

No, but even if they were the magician never sees the card you have taken.

> **The magician changes your card by sleight of hand**

No, because the magician never touches any of the cards.

> **Some other method you don't know about**

Since you can see the trick performed ('The Unexplained Card Trick' by Peter Altman on YouTube at https://youtu.be/d-8B8c2yuMo), you know that it can be done. You just do not know how. Did he read your mind? Did he foretell the future? Or was it something else? It does not matter—the point is that something has happened but you do not know how it happened.

Let us now make a big leap backwards in time to the origin of the universe. The universe exists because we can see it and are part of it so if it arose rather than having existed always, it

Image 17. *YouTube—The Unexplained Card Trick. © Author.*

must have come from somewhere. Perhaps it was M Theory. Perhaps it was something else. We just do not know. Can you see the parallel with the card trick? There is a huge difference in scale of course but the principle is the same. Something has happened but we do not understand how it happened. That is puzzling to be sure but it does not mean that there is no logical explanation.

We can summarise these two events.

Table A. Card trick

CARD TRICK	ORIGIN OF THE UNIVERSE
Can be seen on YouTube	Can be seen all around us
There has to be a method	There has to be a method
Explanation known to magicians	Explanation currently unknown

How many people seeing the card trick would think that the magician had divine help? Not many. It is just a trick. I do not know how it is done and it is very clever but it is just a trick.

Absolutely right! Faced with something we do not understand we do not immediately reach for explanations that defy the laws of nature; rather we accept that we just do not know the explanation. Just because you do not understand how the card trick was done does not mean that there is no logical explanation. Similarly, just because we do not understand how the universe came into being also does not mean that there is no logical explanation. We just have not worked it out yet.

Or there is the option of saying that the universe has always existed, thus removing the difficult problem of how it arose out of nothing. This creates its own difficulty though since it is not easy to get one's mind around the concept of something having existed forever.

For those of you who are brave enough, you may like to look at this article:

> "Something from Nothing? A Vacuum Can Yield Flashes of Light," *Scientific American*, February 12, 2013. https://www.scientificamerican.com/article/something-from-nothing-vacuum-can-yield-flashes-of-light/

You may also like to look up the Casimir Effect. If you understand it, please do let me know.

THIS IS AN UNANSWERABLE QUESTION

Best Guess Answer

The universe exists so either it had to arise somehow or it had to always have existed. It obeys the known laws of physics now so it is a reasonable assumption that it has always done so, even before it appeared (although the laws it obeyed then might not be the same ones we observe today). Under the right conditions, it had to appear, just like, under the right conditions, a car has to run off the road if it goes too fast around a bend. Introducing an additional process, a Creator, is unnecessary since this does not help us to understand the process.

If it has always existed, then there is no need to consider how it arose because it did not have to—it has always been there.

My best guess answer therefore is actually two answers since there is no way of deciding which is better.

The universe either arose on its own as a consequence of the laws of physics and the values of the physical constants by a mechanism currently unknown, or it has always existed.

Q2

DO OTHER UNIVERSES EXIST?
(and what is the weight of an orange?)

What is the speed of light? It is close to 186,000 miles per second, or 300,000 km per second. What is the weight of an orange? We cannot say because it can vary whereas the speed of light through a vacuum is always the same. The speed of light is one of several things in science known as a constant because its value never changes, unlike oranges, for example,

Image 18. *Train. ©Author.*

which can and do vary. An intriguing question is why do the physical constants, like the speed of light, have the values that they do? And what would happen if they were different?

Let us try and put this into the context of everyday life. Imagine a train as in Image 18.

It runs on rails and goes where the rails go until it gets to a set of points where it can go one of several ways, depending on how the points are set. The majority of the world's railways use the standard gauge for their track. This means that the distance between the inside edges of the track is exactly 1,435 mm or 4ft 8½ in.

Now imagine that the train is approaching a set of points where it is due to take a left fork. However, the track to the left is of the 1,422mm (4ft 8in) gauge. What do you think is going to happen to the 100 tonne locomotive as it careers around the bend? Its wheels are now too far apart to fit onto the track so it will jump the rails to be followed by all the coaches behind. If it has not fallen over by now it will do very soon, resulting in a spectacular train crash.

This is an example of how a very small change in the size of something (less than 1 percent in this particular case) can have a very large effect on a wider scale. In a similar way, very small changes in the size of the physical constants would have a very large effect on the structure of the universe.

All life on Earth is based on carbon. This is because carbon has a unique property among the elements in that it can form very long and stable chains and other shapes of molecules needed for life to exist. However, if the values of some of the physical constants involved in atomic structure were only a little bit different, then carbon would not have its special properties and life, at least our type of life, could not have developed. Even more fundamentally, small changes in the masses of protons and neutrons would have changed the likelihood of star formation—no stars, no life. Why then do the physical constants have the values that they do?

There is no good answer to this question. Some would say that the universe must have been designed the way it is, otherwise it does seem like a very fortunate coincidence indeed that the laws of physics are what they are and that we can be here. It is called the fine-tuned universe. As with many things though, there are alternatives. Here are two:

THE VALUES OF THE PHYSICAL CONSTANTS ARE INTERRELATED AND A CHANGE IN ONE RESULTS IN COMPENSATORY CHANGES IN THE OTHERS.

Even though a change in the value of one physical constant could result in a universe incapable of sustaining life, or perhaps existing at all, what if they all change? We saw that a train crash would result if there was a small change in the track gauge after the train reached a set of points. But what if, at the same time, there was also a compensatory change in the spacing of the wheels on the train? The changes compensate for each other and there is no crash. If that is the situation with the physical constants, then it becomes far more likely for them to have values consistent with a life-sustaining stable universe since it is no longer a requirement that the values are precisely what they are now. They could all be different but in a compensatory way. This then implies that the values are somehow related—that a change in one inevitably leads to a compensatory change in the others. It would be interesting to know if there is any evidence to support this notion.

What has been discovered in computer modelling experiments however is that star formation could still occur with many different values for some of the physical constants as long as the changes are compensatory. An article by Fred C. Adams with the intriguing title, "Stars in Other Universes: Stellar Structure with Different Fundamental Constants", which appeared in the *Journal of Cosmology and Astroparticle Physics* in August 2008 explores this.

OTHER UNIVERSES WITH DIFFERENT VALUES FOR THE PHYSICAL CONSTANTS EXIST—THE MULTIVERSE CONCEPT

What if ours is not the only universe? What if there are huge numbers of other universes each with different values for the physical constants and the one that we are in is the one with values of the physical constants that are compatible with life? We are in that one because it is the only one we could be in. This is known as the Anthropic Principle. It is really a philosophical argument that for the universe to be observed by living beings, its conditions must be such that the living beings can exist to do the observing.

Think of a pack of cards. When new, they are usually in factory order arranged numerically and in suits. The total of 52 cards can be arranged in 52 x 51 x 50 x 49 x……….3 x 2 x1 different ways. This is a truly massive number, approximately equal to 10^{68}.

Imagine now that 'factory order' represents our universe with its life-sustaining values of the physical constants and all the other 10^{68}-1 'non-factory order' arrangements represent unstable non-viable universes with other values for the physical constants.

We are not lucky to be in the right one—it is the only one

we could be in.

Finally, we should consider a more fundamental question about the physical constants which is whether they are all constant with time and also with the variety of environments in space. As we have seen, even a small change would lead to a very different universe so on this basis alone we would surmise that the constants really are constant. However, experimental work has been done on this which supports their constancy.

An interesting article ("Are the Constants of Physics Constant?" *Scientific American*, 7 March 2016. https://blogs.scientificamerican.com/guest-blog/are-the-constants-of-physics-constant/) describes work on the mass ratio. This is the ratio of the mass of a proton to the mass of an electron. Being a ratio of two masses, it has no units and its currently accepted best value is 1836.5267389 with an uncertainty of 17 in the last 2 digits. The article cites work from various sources which have measured the ratio by analysing light from astronomical origins billions of light years away. Work has also been done with Hubble telescopic observations of white dwarf stars in environments with very high gravitational forces, up to 10,000 times stronger than on Earth.

The results seem clear, and show that the value of the proton: electron mass ratio is the same wherever one looks. If it is affected by time (over billions of years) and/or by strong gravitational forces, then the effects are extremely small, probably of the order of a few parts per billion.

Many articles have been published on the topic of the constancy of the constants with a variety of results (very small increases; very small decreases; no change).

It must be remembered however that this work is at the extremes of reliability of the technology and that any result would need considerable confirmation from other work before being accepted as true. All we can say at the moment is that if there has been any change to these values over time, then the changes are likely to have been extremely small.

We must bear in mind, of course, that if the physical constants are indeed interrelated as just discussed, then a multiverse could contain vast numbers of life-sustaining universes with different values for the physical constants but ones which have had compensatory changes. Cosmological equations show that there could be as many as 10^{500} universes each with different physical laws.

This concept of multiple universes is not easy to accept or visualise. If we wish to try and form a mental image of these universes we could perhaps imagine a number of floating balloons. Each is different but unlike the picture below they should be imagined as being separated by vast distances with no possible communication between them.

Where are they? Can we see them? Where do they come from? Are they here now? These really are unanswerable questions.

Image 19. Balloons representing separate universes in a multiverse. ©Author.

THIS IS AN UNANSWERABLE QUESTION

Best Guess Answer

We have to try and explain why the laws of physics (that is, the values of the physical constants) are what they are, thereby enabling life to flourish in a stable universe. There has to be a reason the values are what they are. There are various options:
1. The values were fine-tuned by a designer or creator so that life could exist.
2. The values are interrelated so that a change in one triggers a compensatory change in the others. There would therefore be very many possible value combinations consistent with a stable life-sustaining universe. This makes a universe with the 'correct' values for life to exist far more likely.
3. Huge numbers of other universes exist or have existed, each with different values for the physical constants. We are in the one that has the 'correct' values for life to exist.

The best guess answer depends on our beliefs. My personal opinion is for 2 or 3.

Q3

HOW DID LIFE BEGIN?

Life is all around us and within us, so we can be sure that it exists. There was a time when it did not. Depending on your beliefs, this may have been thousands or billions of years ago but for now this does not matter. What does matter is that at some time it did not exist and then it did. So, it had to arise somehow. We can all agree on that.

What are the possibilities? Let us start with these:

It arose on Earth;
It arose elsewhere and was brought to Earth on an asteroid or meteorite.

It has been suggested by some people that life did not arise on Earth at all but was brought here in a primitive form by a meteorite or asteroid collision, and then developed here. This is an interesting theory (known as panspermia) but extremely difficult to prove.

Also, it does not help, even if it were true, because then we would have to ask how it arose in its original location. All this achieves is to place the origin of life somewhere else; it tells us nothing about how it arose.

It is interesting to note, however, that many of the building blocks of proteins, known as amino acids, have been found in comets and meteorites, suggesting that these compounds, which are essential for our type of life, are quite common in space.

There is another interesting thing. Compounds found in life, such as proteins and carbohydrates, are complex organic molecules.

These molecules can exist in two forms—they can be right-handed or left-handed (Image 20). A good analogy is a pair of gloves. Like the molecules, they are mirror images of each other.

For example, we can have d-glucose (d for *dextro*, which means 'right') and l-glucose (l for *laevo*, which means 'left'). It turns out that every glucose and other sugar-type molecule found in living creatures (on Earth) is of the d type. Conversely, all the amino acids (building blocks of proteins) found in living creatures are of the l form. If such a compound were to be made synthetically in a laboratory, it would consist of equal amounts of both the d and the l forms. So how did these two forms become separated and how did only one go on to become the chosen one for building a living creature? We do not know. And what is really interesting is that when these compounds were discovered in some meteorites and comets, as stated above, it was found that there were more of the d sugars than of the l sugars, and more of the l amino acids than the d amino acids.

This in itself does not prove that life came from space since there could be many reasons why right-handed sugars and left-handed amino acids arose on Earth. Cosmic rays, polarised light and lightning strikes could all have had an

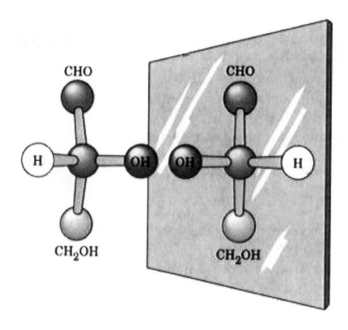

Image 20. *Structural diagram of a glyceraldehyde molecule (a carbohydrate) and its mirror image representing the d and l forms. The diagram shows that in three dimensions the two forms are different, like a pair of gloves. They will therefore have different properties. ©Ali Ramadan, with permission.*

influence on how these compounds arose in the first place. But it is interesting nevertheless.

According to current geological estimates, Earth is about 4.5 billion years old (see Question 4 for more on this). The earliest life forms are thought to have arisen about 3.5 billion years ago, leaving about 1 billion years for the process to have occurred. We have no way of knowing whether that is fast or slow since there are no life forms from other planets that we know about for us to study and compare.

Let us now return to the question of how life might have arisen. In the absence of any real evidence to the contrary, we will assume that this took place on Earth.

So, what are our options? There are only two:
1. Life came into being on its own
2. Life was the work of a Creator

LIFE CAME INTO BEING ON ITS OWN

Someone who believes in a creator or designer will say that this is impossible, pointing out that an object such as a house or a camera or a watch is far less complex than a living creature, and no one would suggest that houses or cameras or watches could just appear on their own. So how could life?

Well, on that basis it could not. But this is a fallacious argument since the analogy is wrong. Obviously, a house could never appear on its own. You could wait forever for this to happen, but no one is suggesting this.

House　　　　　　　　　　　　　　　　*Camera*

Watch movement

A house could not appear on its own, but what about a brick—a random piece of rock that happens to have become shaped in such a way that it could fit next to a similar brick? A few bricks like this and you are beginning to get a wall. (This example is given only to illustrate the point of a gradual development rather than of the finished article arriving in one step. It is not meant to suggest that a house could eventually be constructed in this way, which it could not).

The mistake that many people make when considering and then rejecting the possibility of the emergence of life on its own is to assume that the finished article is formed straightaway. There is nothing and then suddenly there is a crocodile. Obviously, that is not going to happen. It is a gradual process where each step has to be self-sufficient and have an advantage over previous steps. When thought of in this way it becomes feasible.

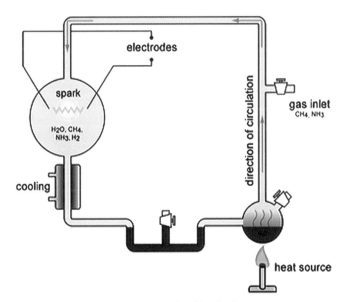

Image 22. The Miller–Urey experiment. ©Wikimedia Commons.

In 1952, two American biochemists, Harold Miller and Stanley Urey, set up an experiment (Image 22). They sealed some of the chemicals thought to be present in the atmosphere of Earth before life had arisen (methane, ammonia, hydrogen and water) in a sterile flask and passed electric sparks through it for a week to simulate lightning strikes.

Amazingly, when the brew was analysed, it was found to contain many organic compounds including amino acids, the building blocks of proteins.

The experiment inspired other scientists to repeat the work, and these showed that some of the components of nucleic acids, the building blocks of DNA, could also be produced in this way. Look up the Miller–Urey experiment for more details.

What this meant was that by starting only with some of the simple chemicals thought to be present on Earth before life began and passing electric sparks through them, it is possible to create many of the more complex compounds needed for life. It is a bit like the brick—it is not a house but it could be the beginnings of a wall. Once the basic building blocks exist, it is at least possible to imagine them coming together to form more complex molecules and eventually small single cells.

You may wonder how and why an electric current could produce amino acids and nucleic acids in this way. Why does your car want to leave the road if you drive too fast around a bend? It is centrifugal force, or, if you prefer, the properties of bodies in motion and the influence of gravity. In other words, things, whether they are molecules or large objects, have certain inherent ways of behaving in certain surroundings. No one 'makes' them behave in this way; they just do.

So it is with a mixture of methane, hydrogen, ammonia

Image 23. Reconstructed vase. ©Author.

and water. When assaulted by electric sparks, these substances rearrange their atoms to form other compounds. It involves the laws of physics or, in this case, chemistry.

Nevertheless, you may think that it was quite a convenient coincidence that the chemical compounds needed to build proteins and DNA just happened to be those that could be produced by lightning strikes and cosmic ray bombardment of Earth's early atmosphere. It sounds a bit like finding some random pieces of rock that just happen to fit together to form a beautiful vase (Image 23).

Proteins, DNA and other essential chemicals needed for life to form did not exist at this time. The laws of physics and chemistry dictated that the chemicals needed for biochemistry to develop could be formed spontaneously as a result of natural processes on Earth, just like they did in Miller and Urey's experiment. If this was not possible, then life would not have developed and Earth would have remained a barren rocky planet.

Using the vase analogy, we can imagine searching amongst random rock chippings in a quarry and eventually collecting some that could be formed into a vase. It may be unlikely but with luck and enough time it is theoretically possible.

So, was the formation of proteins and DNA just a lucky accident? Of course it was, but it only had to happen once, and it took about 1 billion years. It is the same with the random pieces of rock. Most will not form a vase but if you have got a billion years you may just find enough that do.

Going back to the origin of life, it is encouraging that the basic building blocks could have been produced by natural processes on the young Earth. However, the fundamental property of life is that it can reproduce itself. If it could not do that it would not survive.

One more thing needs to be said. The concept of left- and right-handed molecules was mentioned earlier in this chapter. All complex organic substances can exist in both forms but only one of them is able to work in a living cell. The amino acids produced in Miller and Urey's experiment consisted of a mixture of left- and right-handed forms (known as a racemic mixture) but only left-handed amino acids are found in life. Therefore, to have been useful, there would have to have been a way for the racemic mixture to become separated into its constituent left- and right-handed forms.

You might think of this in terms of someone who just needs one glove, either left-handed or right-handed, and is presented with a box containing a mixture of left- and right-handed gloves. He or she just reaches in and grabs one.

It is not known either why left-handed amino acids were

selected instead of right-handed ones, or, in the case of carbohydrates, the right-handed ones over left-handed ones. It is also not known how the separation occurred.

To start with then, what we need is a molecule that can produce a copy of itself. Once we have achieved that, the rest could follow.

Various theories have been put forward as to how this might have happened. Here is one. Think of some jigsaw pieces floating around in a pond. Shaken about, some of them might interlock with each other to form a short chain. To help with the mental image, assume that these pieces are a mixture of pale grey and dark grey, as shown in the illustration (Image 24).

The dark grey pieces might then interlock with the pale grey pieces to form a second chain, so that we now have a pale grey chain and a dark grey chain linked together. If the joining parts of the dark grey chain match those of the

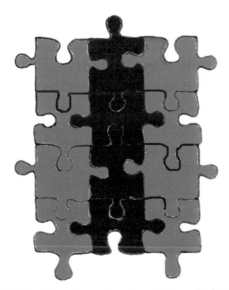

Image 24. *Possible mechanism for self-replicating molecules.* ©*Author.*

pale grey chain, and they come into contact with more pale grey pieces, then a second identical pale grey chain could be formed.

So, what has happened here is that the pale grey chain has reproduced itself *with its constituent pieces in the same order as in the original chain*. This is basically how DNA reproduces itself.

Rather than a pond, as mentioned above, it is now thought that such a process might have occurred in what are known as *hydrothermal vents*. These are fissures, or openings, in the sea bed where the water can be heated to high temperatures by geological activity, thus providing the energy for primitive biochemical reactions to take place. As mentioned elsewhere, this implies that the habitable or Goldilocks zone—that region around a star where a planet could have liquid water on its surface and usually cited as a requirement for life— might need to be modified to take hydrothermal vents into consideration. The icy moons of Jupiter and Saturn, for example, while well outside the habitable zone, might still have liquid water deep under their frozen oceans where life could have arisen.

The function of DNA in cells is to produce the proteins necessary for life to exist. The most important of these proteins are the enzymes (catalysts) that enable all the other chemicals needed by living cells to be synthesised.

Random mistakes and mutations do occur and a small number of these will result in a benefit. A well-known example is the ability of bacteria to become resistant to antibiotics—a big problem nowadays. Bacteria divide and reproduce very rapidly; a single bacterium might divide every 20 minutes so that after 8 hours there would be about 8 million of them! Assuming an average bacterial mutation rate, there could be thousands of mutations in the 8 million bacteria. It needs only one of these mutations to confer some benefit onto the bacterium that could make it resistant to an antibiotic. This single bacterium would then produce another 8 million after another 8 hours, all of which would then be resistant to the antibiotic. The bacterium has therefore evolved into a modified form which is superior to the original and has a greater chance of survival. This, fundamentally, is Darwin's principle of natural selection.

Returning now to DNA, we know that the building blocks of DNA (the jigsaw pieces in the illustration) can be produced by natural processes. So the described scenario is theoretically possible.

Our conclusion, therefore, is that it is possible for the building blocks of life to form on their own, given enough time and suitable conditions. If they were to encounter some fatty or oily materials, then perhaps a wall or membrane could form around them. This would then resemble a simple cell which would have the ability to make copies of itself.

We should also mention the problem of the water on Earth which covers about 70 percent of its surface. It is clearly essential for life but how did it get here? The primitive Earth solidified from a hot molten mass during which time any liquid water would presumably all have boiled off. The prevailing thinking is that it was brought here on a colliding asteroid or comet, an unlikely (see also Question 5) but fortunate occurrence.

What is the likely timeline of all these processes? Here is the current opinion.

ORIGIN OF LIFE TIMELINE

The Earth cooled from its original molten state into a solid rocky planet about 4.5 billion years ago. Fossil and other evidence has enabled geologists and biologists to construct a timeline of how life may have originated. The following is a brief summary.

Table B. Origin of life timeline

TIME	LIFE
4.5 billion years ago	Solid planet Earth. No biology or life.
3.5 billion years ago	Primitive single cells known as prokaryotic cells, with no internal structures. *Example: bacteria*
2 billion years ago	Larger complex single cells known as eukaryotic cells, with internal structures. *Example: amoeba*
1 billion years ago	First multi-cellular organisms. *Example: algae*
200,000 years ago	Homo sapiens

All life on Earth has the same biochemistry and, as far as we know, has arisen only once. If there were other types of primitive life, then we have no record or knowledge of them. Every living thing that exists today, and had existed before, is and was descended from what is known as IDA—the Initial Darwinian Ancestor. IDA led to LUCA—the Last Universal Common Ancestor. Both may have been molecules that had the ability to store information and to reproduce themselves.

Geological and other evidence shows that what we would call 'life' took about 1 billion years to get going. We have seen how a molecule that can make copies of itself could have arisen, this then enabling the primitive prokaryotic cells to reproduce themselves. No significant changes occurred for the next 1.5 billion years, indicating the great difficulty in advancing from a small simple cell to a larger more complex one containing internal structures.

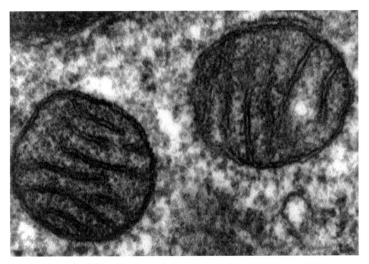

Image 25. *Two mitochondria in a cell from the lung. Magnification c. 50,000 times.* ©*Wikimedia Commons.*

This important step in evolution may have arisen through a chance encounter between two bacteria, where a larger one engulfed a smaller one. The evidence for this comes from a study of the internal structure of complex cells, all of which contain smaller structures inside them. One of these structures, known as mitochondria, resemble miniature cells as shown in Image 25.

Among other functions, mitochondria produce most of the energy required by the cell so the acquisition of these structures by simple prokaryotic cells immediately enabled them to undertake more energy-demanding functions. It then took another billion years for these eukaryotic cells to combine and create the first multi-cellular organisms.

The driving force behind these changes over time—evolution, which actually just means change with time—resulted in the multitude of living things that have and still populate the Earth.

A well-known and excellent example of evolution in action is the peppered moth.

Peppered moths, as their name implies, usually have white with black speckles across the wings. This makes them well camouflaged against lichen-covered tree trunks, protecting them from predators (Image 26). There is also a naturally occurring genetic mutation which causes some moths to have almost black wings, caused by extra production of melanin. These black forms are not as well camouflaged on the lichen as the normal peppered forms and so they are more likely to be eaten by birds and other predators. This means that fewer black forms survive to breed and so they are less common in the population than the paler peppered forms.

In the 19th century it was noticed that in towns and cities it was actually the black form that was more common than the pale peppered form. Industrialisation had caused sooty air pollution which had killed off lichens and blackened many urban tree trunks and walls. So now it was the pale form of the moth that was more obvious to predators, while the black form was better camouflaged and more likely to survive and produce offspring. As a result, over successive generations, the black moths came to outnumber the pale forms. Since

Image 26. *Pale and black peppered moths on pale and dark backgrounds demonstrating the camouflage effect. Look carefully and you will see a black moth to the left of the pale one on the left, and a pale moth below the black one in the photograph on the right. ©Wikipedia.*

moths have short lives, this evolution by natural selection happened quite quickly, actually over a period of just a few decades

More details and photographs are widely available online and in textbooks.

But having said all this, we still have not tackled the problem of how a complex living creature could arise all by itself. The previous examples of houses, cameras and watches made the point that such objects could never arise on their own. So how could objects such as eyes, for example, which are only a small part of complex creatures?

It is a good question until one bears in mind that no one is suggesting that an eye appears out of nowhere. The whole point of evolution is in the name—change with time. Eyes would have started forming with chemicals that react to light, then developing into a structure that contains such chemicals.

Such primitive 'eyes' might only be able to perceive a difference between light and no light, such as is the case with planarian flat worms, but further gradual changes would improve this ability. Each stage of the evolving process works to some degree, and each subsequent stage is then an improvement on the earlier one.

There is substantial literature on this topic which is well worth reading. Look up 'Evolution of the Eye' for further information.

Is it possible that life could have come into being on its own?

Yes, it is possible.

LIFE WAS THE WORK OF A CREATOR

Genesis 1:1 In the beginning God created the heaven and the Earth.

Genesis 1:11 And God said Let the Earth bring forth grass, the herb yielding seed.

Genesis 2: 2 And on the seventh day God ended his work which he had made.

Scientific advances during the 19th century led many people to abandon the literal interpretation of the Bible and to accept the view that the Earth and universe were very old. By the early 20th century, Darwin's ideas concerning evolution were also becoming widely accepted.

In 1920s' America, a Christian fundamentalist movement began which was opposed to the idea of evolution and succeeded in getting its teaching banned in schools. The famous Scopes Monkey trial of 1925, later made into the film *Inherit the Wind* (1960), resulted in a school teacher being fined for teaching evolution to his class.

The term Creationism was coined as an alternative model to that proposed by evolution. In the 1960s, this became Scientific Creationism although still proposing a literal interpretation of the Book of Genesis. In the 1990s, this then became known as Intelligent Design to strip it of any biblical references. It holds that intelligent intervention was necessary for the creation of life and of the universe.

Today, a very large number of people of different religions still interpret the writings of the Bible literally, and therefore believe that the entire process of creation did, in fact, take just six days.

That is what it says in the book. But it is an old book. You have to ask yourself, how many 3,000-year-old ideas are still true today? Just because people thought something was true in the past does not mean that you have to believe it now.

Would you want your doctor to be using medicines from the Ancient Egyptian era? Or would you want your dental surgeon to be removing some teeth with 14th century techniques and equipment (image 27).

Is the Earth flat? Do the Sun, Moon, planets and stars all revolve around the Earth? If you are buried with all your earthly possessions, can you use them in an afterlife? Is everything made of earth, air, fire and water?

The belief is that life and the universe were the work of a creator or designer. Some adherents believe in a literal interpretation of the biblical timeline and hold that the universe is only about 6,000 years old. Others, while still believing in a creator or designer God, do accept the scientific evidence for a very old universe.

Image 27. *Dentistry in the 14th century (from the* Omne Bonum, *a 14th century encyclopaedia).* © *British Library.*

We will come back to the age of the Earth and universe in Question 4 but for now we are concerned with the Creator option for the creation of life.

We have seen that scientific evidence does exist to allow for the possibility that life arose on its own, this being a slow and gradual formation from simple self-reproducing compounds to simple cells and eventually to more complex organisms. To reiterate, no one is suggesting that finished complex living things suddenly appeared on their own.

It is possible that life arose on its own. It probably took about 1 billion years to get started and it may have been very unlikely but it only had to happen once. In that case, why complicate the explanation by saying that a Creator, about whom we know nothing, did it? We have to admit, however, that even though plausible mechanisms for the origin of life can be put forward, how it actually happened is a question we cannot answer. All we do know is that it happened, because here we all are! And invoking Ockham's Razor, we can conclude that the simplest explanation, with no Creator involvement, is likely to be the best explanation. (But see also Question 14).

THIS IS AN UNANSWERABLE QUESTION

Best Guess Answer

It seems most likely that life arose on Earth after the basic building blocks (amino acids, nucleic acid components and carbohydrates) were produced as a result of lightning strikes and cosmic ray bombardment of the primitive Earth's atmosphere. These components came together by random collisions in mixtures, possibly in deep hydrothermal vents in the ocean, and formed simple compounds capable of self-replication, eventually forming simple cells. Complex cells with internal structures exist in most life forms and probably arose by the chance ingestion of a small cell by a larger one. The process of natural selection or evolution then took over eventually resulting in the enormous variety of living things seen today—around eight million different species.

In this scenario, the vast timescale available is thought to be sufficient for these processes to have occurred naturally.

Q4

HOW OLD ARE THE EARTH AND THE UNIVERSE?

Estimates of the age of the Earth have increased steadily over the past several hundred years from the biblical value of about 6,000 years to the currently accepted scientific value of about 4.5 billion years. Having said that however, there remain many people who cling to the biblical version of events and maintain that the Earth is only 6,000 years old.

We therefore have two options—the scientific and the biblical.

THE SCIENTIFIC OPTION

Creationism and Intelligent Design are fundamentally the same thing with different names. While all proponents of these concepts believe in a universe created by a God, they differ in other aspects. Some maintain that the universe is as old as estimated by astronomers while others say that the biblical account is literally true, that creation did take just six days, that God created each kind of life individually, and finally, that it is no more than 6,000 years old.

If this is the case, then how is it that we can see stars and galaxies that are millions and billions of light years away? Imagine that we are looking at a galaxy that is 50 million light years away. What this means is that the light from that galaxy takes 50 million years to reach us. If we can see it, then it and the universe must be at least 50 million years old. The furthest galaxy currently known is over 13 billion light years away, so the universe must be at least 13 billion years old. How do Creationists get around this problem?

It is called in-transit creation and states that God created this light in transit about 6,000 light years away. This is then consistent with the biblical age of the universe and also explains the apparent much older age measured by astronomers.

It also means that events seen and measured by astronomers, such as supernovae, other galaxies, black holes, quasars, etc., do not really exist since the light (and radio waves and X-rays) by which we see and detect them

Image 28. *Andromeda galaxy. ©NASA.*

were placed in space by God 6,000 years ago to give us the impression of a much older and more complex universe.

Image 28 is a photograph of the Andromeda spiral galaxy. It is 2.5 million light years away and on a clear night is just about visible to the naked eye as a faint smudge. Do you believe that it does not really exist and that its light was placed in space by God to fool you? Why? What would be the point? If these objects do not exist, then why invent them? It makes no sense.

Furthermore, all of the very old objects in the sky are so far away that they can only be seen with powerful modern telescopes. Why invent something that no one could see until telescopes were invented?

The case for a young Earth arises from a literal acceptance of the Bible, in particular, those portions of the Old Testament which provide an unbroken male lineage from Adam to Solomon, complete with the ages of the individuals.

Further analysis of events detailed in the Old Testament has resulted in a chronology from creation up to the birth of Jesus.

Several such chronologies have been published, and one of the best known is that of James Ussher, a 17th-century Irish Archbishop. He provided an exact date for creation—23 October 4004 BCE. This value is so different from the scientific one of about 13.7 billion years for the universe, and about 4.5 billion years for the Earth, that it should be easy to prove one way or the other.

Here are a few arguments for a very old Earth and universe:
- Dating of rocks by radioactive decay
- Ice core and tree ring data
- Observance of galaxies billions of light years away

These arguments seem irrefutable. Radio isotope dating is a well-established scientific dating method; counting ice core layers and tree rings just needs time and care, and being able to see galaxies billions of light years away means that their light must have started the journey many billions of years ago to get here for us to see it now.

However, all these seemingly solid arguments can easily be put aside. The young Earth creationists' response depends on two rebuttals.

RADIOACTIVE DECAY

The rate of radioactive decay we measure today may not be the same as it was in the past, thus giving false results. Since no one was around to measure these parameters millions or billions of years ago, it is not possible to say with absolute certainty that these rates were the same then as they are now. If they have in fact changed, then the calculated age of the Earth could be 6,000 years rather than 4.5 billion years.

This is a clever, and to be fair, reasonable argument. To answer this, we first need to describe in some detail the process of radioactive decay and the concept of half-lives.

Image 29 shows six sand timers. Each one runs for a different number of years as indicated. The largest one on the left takes 1 billion years for all of the sand to fall to the bottom. The next one takes 1 million years, and then 100,000 years, 10,000 years, 1,000 years and, finally, the smallest one on the right takes 100 years.

Now imagine that you come across these timers standing in a row as in the diagram. You will see that the 100-year and the 1,000-year ones are finished, and that the 10,000-year one is soon to finish. The 100,000-year one and the 1 million-year one both still have a considerable time to go

Image 29. Sand timers with different running times to illustrate the principle of radioactive dating. ©Author.

while the 1 billion-year one has barely changed. What is your conclusion about how long ago they were set up?

It is pretty obvious that the answer is just under 10,000 years. The shorter ones have finished, and the longer ones have not.

This example is meant to illustrate, in a simplified way, how the radioactive decay of elements can be used to date the Earth. First though, we need a brief description of radioactive decay.

Elements are made of atoms, and each atom consists of electrons spinning around a nucleus made of protons and neutrons.

Image 30 is a simplified drawing of a carbon atom. There is a central nucleus that contains six protons (black circles) and six neutrons (white circles). Spinning around the nucleus are six electrons (black dots). It is a bit like a mini Solar System with the Sun at the centre and the planets spinning around it.

Many atoms exist in multiple forms called isotopes. These differ from each other in the numbers of neutrons in the nucleus. Carbon 14, for example, has eight neutrons rather

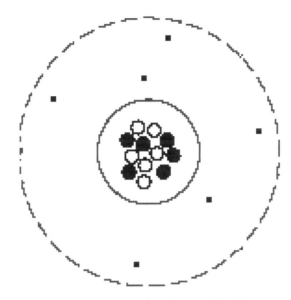

Image 30. Diagram of a carbon atom. ©Author.

than six (eight neutrons plus six protons = 14 particles in total, hence carbon 14).

Some isotopes can be unstable and over time will lose some of their electrons, protons or neutrons. This then results in the original atom, say uranium, for example, changing into another one (which in the case of uranium would be thorium).

This process is known as radioactivity and the change from one element into another is known as radioactive decay. Some isotopes are stable in that they do not decay at all. However, in 2003, French scientists found that an isotope of the element bismuth, previously thought to be stable, did in fact decay but so slowly that its half-life (see below) was 20,000 quadrillion (2×10^{19} years). Conversely, some elements are so unstable that their half-lives are measured in yoctoseconds (10^{-24} seconds). The most unstable, according to a Wikipedia article, is hydrogen-7 which has a half-life of 23 yoctoseconds.

We are not used to objects being unstable in everyday life so the concept may not be that easy to visualise. A book, a telephone, or a pair of shoes do not change unless they become damaged.

However, think of a house of cards. It is likely to be rather unstable and if we try and build another floor it might collapse as shown in Image 31. That is a reasonable way of imagining an unstable atom. It contains more particles—protons, neutrons and electrons—than it can reasonably hold together, and some of them escape.

We now need to explain the term half-life. This means the average time taken for half of a radioactive substance to decay into another substance. As just discussed, half-lives vary a lot.

Other examples include the half-life of uranium 238

Image 31. *Collapsing house of cards. © Getty Images, with permission.*

which is about 4 billion years, while that of nobelium 248 is two-millionths of a second. Half-lives are used by geologists to determine the ages of rocks and fossils, and to date other artefacts such as the Turin Shroud. Some information as to how they are used is given here.

Imagine that a snooker ball manufacturer finds that all their coloured balls (shown as grey in Images 33, 34 and 35) have a tendency to fade into white balls and customers have been complaining about this. The manufacturer wants to be able to assess how long ago the balls were manufactured. (Yes, he could look at the paperwork but this is just an example to illustrate the principle). They therefore make tests on large batches of new balls and find that, on an average, half of a batch has turned white after ten weeks. In technical terms, this means that the half-life of the colour change to white is ten weeks, usually written as T½ = 10 weeks. (Note that it does not matter how many balls are in the sample—the half-life would be the same whether a sample contains 100, 10,000 or 1 million balls. Clearly though, a small sample of just a few

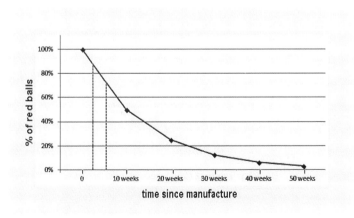

Image 32. *Graph showing time against faded colour of snooker balls. ©Author.*

balls would not be expected to give an accurate result.

In the graph, newly-manufactured balls (time zero) are all grey. After ten weeks, 50 percent have turned white. After another ten weeks, that is a total of 20 weeks, another 50 percent have turned white leaving 25 percent grey (50 percent of 50 percent). Similarly, after a total of 30 weeks only 12.5 percent remain grey and so on. In other words, half of the remaining grey balls change into white after each ten-week period. This creates a sloping graph which eventually tails off to zero percent grey. This is known as exponential decay.

Now look at Image 34 which shows that two balls out of 20 have changed from grey to white. That is a change of 10 percent meaning that the percentage of grey balls remaining is now 90 percent. Reading from the graph, we can work out that this batch of balls is two weeks old. In the same way, we know that the batch in Image 35, which has 75 percent of grey balls remaining, is five weeks old.

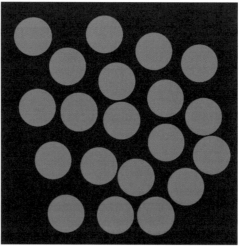

Image 33. *All balls are grey. ©Author.*

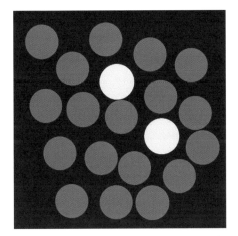

Image 34. *Ten percent have faded to white. ©Author.*

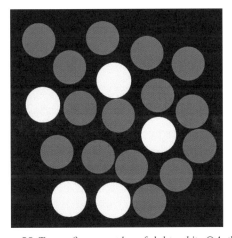

Image 35. *Twenty five percent have faded to white. ©Author.*

This is how half-lives are used to date ancient items. Instead of fading colours, geologists use decaying radioactive elements but the principle is the same.

A commonly used substance for dating rocks is potassium-40, a radioactive isotope of potassium. This is an unstable material and it slowly decays (changes) into argon-40 with a half-life of about 1.25 billion years. Investigators would look at what percentage of potassium-40 remains in a rock and then calculate its age much like the snooker ball example above.

This method does rely on the assumption that when the rocks formed (solidified from their original molten state), any argon-40 (being a gas) would have escaped from the molten rocks but would be trapped once they became solid. In other words, the concentration of argon-40 at the time the rocks became solid is taken as zero. Unlike the snooker balls, which can be seen to be grey at time zero, this is an assumption since the ancient rocks were obviously not examined when they were formed.

There are other, more complex, dating methods which do not rely on making such assumptions. Isochron dating relies on making additional measurements of a second decay product which is itself stable.

(Look up half-lives, exponential decay, radiometric dating and isochron dating for further information).

The 'half-life' is a better measure to use than the 'whole-life' because the latter obviously depends on how much material there is to start with. The half-life, on the other hand, is independent of this.

Going back to the snooker ball example, we need to know how long it takes for half of a batch to fade to white. The size of the batch, as long as it is not too small, does not matter. If we were to use the 'whole-life' then we would have to wait for all of the batch to fade and this would depend on the size of the batch.

Radioactive decay is very useful for geologists as it gives them a whole set of sand timers with different running times.

They would have been set up when the Earth became a solid planet and the rocks stabilized from their original molten state.

If half of something decays (disappears) in 1 million years, for example, then after 2 million years another half will have gone and only one quarter will be left (half of a half), and so on. After ten half-lives (10 million years) only one-thousandth (actually 1/1024) of the original substance will be left. Eventually, a substance will become undetectable since there will only be such a minute amount of it left that it will not be possible to measure it.

Table C on page 69 shows some radioactive isotopes with their half-lives. Any found on Earth are the remnants of what was originally present when the Earth was formed (plus, in some cases, those isotopes formed subsequently by cosmic ray bombardment of the atmosphere). Those not found on the Earth have half-lives so short (in cosmological timescales) that although they would have been present at Earth's formation, they have now all either decayed completely or remain in such low concentrations that they are unmeasurable.

You will notice that the isotopes with half-lives greater than about 100 million years can be detected on Earth whereas those with half-lives less than this figure cannot be detected. (Plutonium 244, with a half-life of 82 million years, has been detected in very minute quantities because a group of geologists made a supreme effort to see if they could find it. If someone tried just as hard to find samarium 146 they too might be successful).

A reasonable question might be—if the isotopes in the bottom part of the table are not detectable on Earth, then why do they have names and how do we know what their half-lives are? All of these substances can be produced in nuclear

Table C. Some radioactive isotopes and their half-lives

Isotope	Half-life (years)	Found on Earth?
Vanadium 50	6,000,000,000,000,000	yes
Neodymium 144	2,400,000,000,000,000	yes
Hafnium 174	2,000,000,000,000,000	yes
Platinum 192	1,000,000,000,000,000	yes
Indium 115	600,000,000,000,000	yes
Gadolinium 152	110,000,000,000,000	yes
Tellurium 123	12,000,000,000,000	yes
Platinum 190	690,000,000,000	yes
Lanthanum 138	112,000,000,000	yes
Samarium 147	106,000,000,000	yes
Rubidium 87	49,000,000,000	yes
Lutetium 176	35,000,000,000	yes
Thorium 232	14,000,000,000	yes
Uranium 238	4,500,000,000	yes
Potassium 40	1,250,000,000	yes
Uranium 235	704,000,000	yes
Samarium 146	103,000,000	no
Plutonium 244	82,000,000	minute traces
Niobium 92	35,000,000	no
Curium 247	16,000,000	no
Lead 205	15,000,000	no
Hafnium 182	9,000,000	no
Palladium 107	7,000,000	no
Caesium 135	3,000,000	no
Technetium 97	3,000,000	no
Gadolinium 150	2,000,000	no
Zirconium 93	2,000,000	no
Technetium 98	2,000,000	no
Dysprosium 154	1,000,000	no

reactors and other atom-smashing machines and therefore can be studied. They would have existed on Earth at one time but, as explained above, have all decayed into other products and are therefore now undetectable.

We can now return to our original question about the age of the Earth. The scientific age is about 4.5 billion years while the new Earth Creationist age is about 6,000 years. There is a break in the half-life table somewhere in the region of 80 to 100 million years. Isotopes with half-lives longer than this can be detected on Earth whereas isotopes with half-lives shorter than this cannot be detected.

If the Earth was 6,000 years old, then even the last isotope in the table, dysprosium 154 with a half-life of 1 million years, would be abundant since it would not even have been around for one half-life. A 6,000-year-old Earth would contain every isotope in the table, and a great many more with shorter half-lives as well.

On the other hand, a 4.5-billion-year-old Earth would be expected to have depleted its original stock of isotopes with half-lives less than about 100 million years since this represents about 45 half-lives (45 x 100 million years = 4.5 billion years). After this time, there would be only about one 35 trillionth of the original substance left, and this would be undetectable.

The evidence is therefore completely compatible with an Earth age of around 4.5 billion years and completely incompatible with an Earth age of 6,000 years. However, what about the notion that the decay rates have changed or that they are wrong or that the calculations used by physicists and geologists are flawed?

Let us examine this idea. A well-known method for dating trees is to count the annual rings. Ice cores can also be dated by counting the ice layer.

Radioactive dating techniques, dendrochronological (tree ring counting) techniques and ice core counting techniques all give the same answers up to several thousands of years. So here we have two completely different methods based on completely different principles that give the same results.

A good analogy would be a comparison between two different types of scales.

Images 36 & 37. *Two different types of scales. ©Wikipedia.*

The scale on the left is a traditional type of balance device based on the lever principle. When the weights in both pans are the same, the beam is horizontal as indicated by a pointer in the middle of the beam. The weights in the right-hand pan are then equal to the weight of the item in the left-hand pan.

The scale on the right depends on the depression of the base. The base is connected to the dial pointer by a series of springs and connectors; the greater the depression (weight), the more the dial moves. The point is that these two weighing devices are based on completely different principles and therefore are extremely unlikely to suffer from the same errors. If an item weighs the same on both scales we can be confident that this is its true weight.

ICE CORE AND TREE RING DATA

It is the same with the two different methods for measuring the ages of trees and ice cores by tree ring counting and radiocarbon dating. If both methods give the same or very similar results, we can also be confident that the results are correct.

Image 38. Tree rings.
©Wikipedia.

Image 39. Ice core layers.
©Wikipedia.

Image 40 shows the ages of ice cores by counting the layers and by radiocarbon dating. The correlation is good, leaving no doubt that the carbon-14 dating method is accurate to at least 25,000 years.

As was discussed earlier in this chapter, we know that isotopes with half-lives of less than about 100 million years are no longer detectable on Earth because they have been through so many half-lives that the amount left is too small to measure. How could the absence of, say, niobium 92 with a half-life of 35 million years be compatible with a 6,000-year-old Earth?

For a radioactive isotope to become undetectable, it would have to have been through about 40 or 50 half-lives. In 6,000 years, the amount of niobium 92 would hardly have changed at all, and there should be lots of it about. For it to have become undetectable in just 6,000 years, its half-life would have to be about 120 years (because 50 half-lives in 6,000 years implies a half-life of 120 years) rather than 35 million

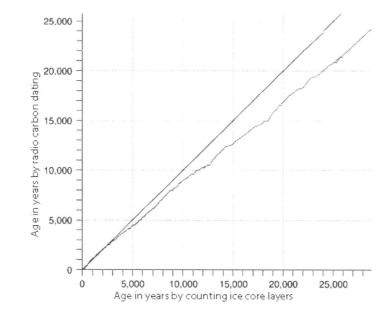

Image 40. *Comparison between carbon-14 and ice core dates (from Greenland Ice Core Chronology 2005).*

years. That is a staggering difference from what scientists have observed. It is a difference of 300,000 times. Similar calculations apply to all the other radioactive isotopes listed in the table that have become undetectable.

It is just inconceivable that all the half-lives that have been measured are wrong by such large factors. Bear in mind that the half-life of carbon 14 is known to be correct since it has been independently calibrated against known ages of trees and ice cores. Why should this be right and so many others be wrong?

We can also look at some rock samples (from Norway) that have been dated by different people using different isotopes, as shown in the following table. They have all given pretty much the same results.

Table D. Ages of rocks

DATING METHOD	CALCULATED AGE
Argon40/Argon39	588 million years
Potassium/Argon	575 million years
Rubidium/Strontium	578 million years
Lead/Lead	573 million years
Thorium/Lead	580 million years

All the samples were from the same rock, and they were dated by using different isotopes, each with different half-lives and different decay mechanisms. Again, it is inconceivable that they have all given the same wrong results (data from Joe Meert (2000); gonwanaresearch.com/radiomet.htm).

Say, for example, you were timing a race and someone said that the time was wrong because your watch was running too fast or too slow. So you check with another judge who was using a quartz watch and who recorded the same time as your mechanical watch. Are you going to say that both timepieces with different mechanisms have the same error so that they all give the same but wrong time? Of course not.

OBSERVANCE OF DISTANT GALAXIES

It is pretty clear, therefore, that radioactive decay is a valid technique for dating rocks and other materials. But what if you still do not believe? What if you still think that the decay rates were different millions and billions of years ago and all the results are wrong. After all, no one was around then to measure the rates.

Actually, that is not quite true. We do have access to a remarkable time machine that can help us here.

When you look at the Moon you are seeing it as it was about one and a half seconds ago since that is how long it takes for the moonlight to travel the 385,000 km (240,000 miles) from the Moon to the Earth. So you are actually looking back in time by one and a half seconds. With the Sun, it is just over eight minutes. And with the Andromeda galaxy, it is 2.5 million years.

Because astronomical objects are so far away, the light by which we see them takes so long to get here that we are looking at these objects as they were in the distant past. Any information that can be extracted by examining the properties of the light from these objects will also be as it was when the light left. In other words, we can examine light that is millions or even billions of years old.

A star can explode for various reasons, and when it does it can form what is called a 'supernova'. This is an extremely bright object and many were observed by ancient astronomers hundreds of years before the telescope was invented.

Complex chemical reactions take place inside a supernova and these result in the formation of many radioactive isotopes. These isotopes can be detected and analysed.

One such supernova that has been studied in great detail is known as SN1987A. It is in a galaxy close to the Milky Way and is 169,000 light years away. That means that the light that we see it by is 169,000 years old. Spectroscopic analysis of this light shows that the decay rates of the radioactive isotopes produced by the supernova 169,000 years ago are the same as those measured here on Earth now. Similar results have been obtained from other supernovae much further away.

CONCLUSION

In summary, we can say that the evidence we have shows no indication that radioactive decay rates have changed over time. This is consistent with the conclusion that the values of the physical constants discussed in Question 2 have also not changed over time, at least to any significant extent.

More precise measurements with larger telescopes will no doubt refine these conclusions in the future but for now it seems safe to say that the current values for these numbers are the same as they have always been.

The concept that the universe may have a finite age rather than having existed forever only arose in the 1920s. As already discussed, the American astronomer Edwin Hubble discovered that galaxies were large accumulations of stars outside and far away from our own Milky Way. He discovered that all the galaxies were speeding away from us and from each other, and the further they were, the faster they were receding.

Calculating backwards, it was possible to arrive at a time when all the galaxies and all the matter in the universe existed as a single point that then exploded in the Big Bang that created the expanding universe we see today. This time is about 13.7 billion years ago. It is consistent with the estimated age of the oldest stars which is about 13 billion years.

It is also important to mention another calculation that gives a similar result. Carbon, which is an essential element for our type of life, is produced along with all the other elements, by reactions inside stars, and is then distributed into space by supernova explosions as some stars die. This is the only source of carbon and without it we would not be here.

From what is known about the reactions that occur inside stars, it has been calculated that this process of forming and distributing carbon (and other heavier elements) throughout the universe takes around 10 billion years. Admittedly, this is not an exact calculation but even the approximation is within the ball park of the scientifically accepted universe age of 13.7 billion years.

To consider that it could have been done within two-millionths of that time period—6,000 years—goes against everything that is known about the reactions that occur inside stars.

THE BIBLICAL OPTION

This option, although seemingly preposterous, is just impossible to refute.

> *God made it seem as if the Earth and universe were very old by putting in place the right amounts of radioactive compounds, tree rings, ice cores and the light of galaxies 'in transit', so that scientific measurements would be fooled into giving incorrect answers, indicating a very old Earth and a very old universe.*

Why? What (on Earth) would be the point? We cannot refute this but we can employ Ockham's Razor and ignore it.

THIS IS AN ANSWERABLE QUESTION

Answer

Dendrochronology shows that some trees are several thousands of years old and these ages correlate exactly with those obtained by radioactive dating of the same tree samples with carbon-14. Radioactive isotopes with half-lives of about 100 million years or less are undetectable on Earth. They would have to have been through about 50 half-lives for this to happen, which equates to a time period of up to 5 billion years. For such isotopes to have become undetectable in 6,000 years, their half-lives would have to be about 120 years which is a staggering error in measurement or change in decay rate.

Also, and very convincingly, different isotopes all give the same dates for the ages of old rocks. Finally, half-lives measured from the light of distant supernovae confirm that the rates measured on Earth now are the same as they were hundreds of thousands and millions of years ago.

Distant galaxies have been measured as being billions of light years away, meaning that the universe must be billions of years old. One of the most distant objects known, which was detected by a NASA satellite on 23 April 2009, is known as GRB 090423 (gamma ray burst 2009 April 23).

The light from this stellar explosion took 13 billion years to reach the Earth and, therefore, started its journey when the star exploded 13 billion years ago.

That means that the universe must be at least 13 billion years old. Calculating back from the known present expansion of the universe actually gives the date of formation of the universe as about 13.7 billion years ago.

James Ussher's and other people's chronologies are based on a literal interpretation of biblical references and an extrapolation of timelines based on quoted ages. For example, in Genesis 5:26 it is stated that "all the days of Methuselah were 969 years, and he died".

The notion that a human being could live for nearly 1,000 years, even in modern times, is unimaginable.

The explanation, presumably, is that either the biblical 'year' is not quite the same as our modern year or that the biblical narratives are incorrect. Whatever the actual reason, the chronologies are invalid.

In January 2020, new studies on the Murchison meteorite, which fell in Australia in 1969, have found that some of the granules inside it are 7 billion years old, that is, 2½ billion years older than the Earth. It is also rich in organic compounds including amino acids and nucleotides, all of which are essential for our type of life. Although most are composed of both left- and right-handed molecules (see Question 3), in some cases there was a preponderance of L amino acids, the form used by living creatures on Earth. This may have implications for life's origin.

The Earth is about 4.5 billion years old and the universe is about 13.7 billion years old.

Q5

DOES ALIEN LIFE EXIST?

Not that many years ago, any talk of aliens would have been firmly in the realm of science fiction. It was a popular genre, with authors such as Isaac Asimov, Ray Bradbury, Nigel Kneale and Arthur C. Clarke, and films and TV shows including *Invasion of the Body Snatchers* (1956), *The Quatermass Experiment* (1955), *2001: A Space Odyssey* (1968) and *Star Wars* (1977) contributing to it. As detailed later in this chapter, the early 1960s saw the beginnings of serious attempts being made to search for radio messages from alien life elsewhere in the universe. Although it had been long suspected that other stars also had planetary systems, this was not confirmed until 1992 with the discovery of the first exoplanet (a planet in another solar system). As of October 2020, 4,292 such planets have been confirmed in 3,185 solar systems. Given the vast number of exoplanets which are likely to exist in the universe, the question of whether any of these are home to alien life has become mainstream science.

One of the really big unanswered questions that regularly dominates the news is whether alien life exists elsewhere in the universe. The prospect of contact with advanced alien civilisations and what we might learn from them is at once an exciting and daunting one. Have they cured disease? Have they found a way for everyone to live in peace? Have they harnessed unlimited sources of energy? The list goes on.

Certainly, some have argued that making contact could be a big mistake since we would not know whether the aliens were friendly or hostile. True, but we are getting away from the basic question as to whether such life exists at all. Since this is a science book, we need to look at the evidence or lack of it. We shall consider the evidence for UFOs, for alien visitations in the past and for actual alien encounters.

I. UFOS

The term UFO (Unidentified Flying Object) was coined by the United States Air Force in 1952 to describe any flying objects that remained unidentified after expert scrutiny. It arose from a need to give a name to an increasing number of sightings of such objects. The first such sighting to be recorded was that of John Martin, a Texan farmer, who described seeing a circular flying object that was about the size of a saucer flying at a 'wonderful speed'. This is believed to be the origin of the term 'flying saucer'. This story was published on 25 January 1878 by the *Denison Daily News*.

There have been many thousands of reported UFO sightings around the world and they continue to be reported both by people on the ground and also by pilots during flights.

Image 41 shows a typical sighting dating from 1967.

So, what are they? Most are probably due to atmospheric or climatic conditions such as unusual cloud formations, unusual lightning patterns, astronomical events, weather or military balloons or other aerial devices, or merely mirages or imaginary sightings. There remains however a large body

Image 41. *UFO sighting in New Mexico, USA.* ©The Paranormal Borderline, with permission.

of people who insist that at least some of the sightings are of real aerial devices with the suggestion that they are of extraterrestrial origin.

Numerous such photographs have been published. The difficulty is that they are so easy to fake that most people attach very little credence to them.

Image 42. *UFO?* ©*Author.*

Image 42 is a photograph that I took a few years ago near my house. It is a bit blurred but there is some structure visible on the object. Now take a look at Image 53.

In 2010, the National Archives of the United Kingdom released into the public domain a large number of previously restricted files on UFO sightings. Included in this archive were many drawings that had been sent in by members of the public. Provided below is a small selection together with brief descriptions of the sightings.

Even this small selection encompasses a wide variety of shapes and designs. The free-floating aliens (Image 44) must have been a remarkable sight for this airline passenger. The Toblerone-shaped object (Image 48) is the length of a London bus and nearly three times as wide. What a pity no one thought to take some photographs as it hovered 3 m (10 ft) above the ground for 40 minutes. Note the helpful

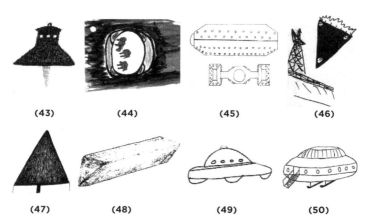

Images 43–50. *UFO drawings submitted by members of the public. ©2010 National Archives with permission. Top row from the left—30-ft wide with flashing blue lights; aliens seen through aircraft window; side and front views and covered in lights; no description given. Bottom row from the left—no description given; 40-ft long and 20-ft wide, hovered 10 ft above the ground for 40 minutes; seen sucking up water; 40-ft wide with steps and sleds.*

stairway and sleds in Image 50. The triangular shapes in images 46 and 47 are reminiscent of the American Stealth bomber as shown in Image 51.

The American authorities would have encouraged members of the public to think that they had seen a UFO rather than a prototype secret airplane.

Perhaps the best-known UFO event was the Roswell incident. This occurred in the summer of 1947 near the town of Roswell in New Mexico, USA, when the Roswell Army Air Field (RAAF) announced that it had recovered debris from a crashed 'flying saucer'. This was reported in the *Roswell Daily Record* (see Image 52). Subsequent reports changed this, stating that the debris was from a high-flying weather balloon.

UFO enthusiasts seized upon this story which soon became embellished with reports of alien corpses being secreted away by the US military. There have been numerous books, TV programmes and films about this event, and although many people accept the balloon interpretation,

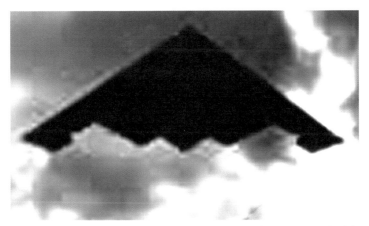

***Image 51**. Drawing of US Air Force Stealth bomber copied from a photograph of the actual airplane. ©Author.*

Mysteries of the Universe

Image 52. Newspaper headline from 8 July 1947. ©Roswell Daily Record, with permission.

Image 53. This is the original photograph from which I created image 42. It is a water butt lid held up on a broom handle. A few minutes of work in Photoshop to remove the handle and blur the image was all that was needed to create a passable photograph of a UFO sighting.

there are also others who are convinced that the US military have concocted a cover story to hide the truth about this event.

It is certain that something fell out of the sky and it is also certain that the US military were conducting secret work with an experimental flying craft with a view to spying on the Soviet Union. It would suit them to support the notion of UFOs to deflect attention from their clandestine activities, and the crash of a secret military device does seem a much more likely explanation that the crash of an alien spaceship. Even so, the story persists although the following should really put an end to it.

In 1995, a film that purported to show an autopsy being performed on an alien creature recovered from the crash site made headline news around the world. Experts in film technology, anatomy and pathology attested to its authenticity.

In 2017, Spyros Melaris, a filmmaker and magician, announced that the autopsy film was a hoax.

I have attended Mr. Melaris's lecture on this subject and it is a fascinating account. Perhaps the most inventive aspect was how he and his co-workers were able to convince

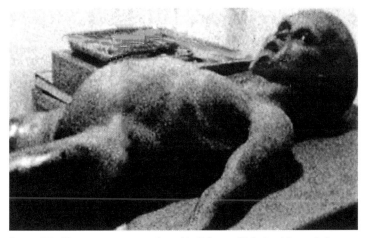

Image 54. *Alien creature on operating table. ©Spyros Melaris with permission.*

Image 55. Original painting of the Bellonzo-Shriever-Miethe flying disc. ©Jim Nichols, 1990.

professional film historians that the autopsy was filmed on genuine 1947 film stock. More details can be found on www.spyrosmelaris.com

Experimental circular aircrafts could have been the source of many UFO sightings in the 1940s and 1950s. These were prototypes to gauge whether this type of craft could be useful as spy planes during the post-War and Cold War periods.

Image 55 is a painting of a prototype German flying disc, sometimes known as the V7, or the Bellonzo-Schriever-Miethe Disc. It reportedly flew in Prague, Czechoslovakia, in 1945, but was never developed further.

Another circular aircraft was the Avro Canada VZ-9 Avrocar (Image 56) was built for the US military in the late 1950s. It never reached its design specifications and after various trials and modifications was retired in 1961. Only two craft were ever built.

Image 56. The Avro Canada VZ-9 Avrocar. ©Wikimedia Commons.

Anyone observing these and possibly other secret aerial devices in flight could easily have assumed that they were real flying saucers.

It turns out though that there are stability issues with this type of craft which is not to say that these could not be overcome.

Unusual cloud formations are probably a very common source of reported sightings. Lenticular clouds are lens-shaped cloud formations that can look strikingly like what many people would call a flying saucer, as shown in Image 57.

Photographs are easy to fake, drawings may or may not represent what was seen, and some sightings might look like strange flying crafts but could be secret military prototype

Image 57. Lenticular cloud formation, Ireland, 2015. ©Wikimedia Commons.

aircrafts or atmospheric phenomena. But let us for a moment imagine that at least some of the sightings are of, in fact, real alien spacecrafts. Two questions arise:

1. WHERE ARE THEY FROM?

Our Solar System

At the present time (2020) spacecraft have visited every planet and many of the moons in our Solar System and found no evidence of any life. A civilisation sufficiently advanced to be able to send spacecraft to Earth would surely have some traces of its existence on its home planet which would have been detected by our spacecrafts' cameras. On present evidence therefore, it is almost certainly safe to assume that there are no alien creatures living in our Solar System who are able to send spacecraft to Earth. Whether some form of microbial or primitive plant life exists is currently the subject of intense research, especially on Mars and Venus, and in the hopefully near future, on some of the moons of Jupiter and Saturn.

Another Solar System

In view of the above conclusion, we have to look further afield to another Solar System. As we have already noted, nearly 4,300 exoplanets have currently been confirmed in over 3,100 solar systems (NASA Exoplanets Archive).

The nearest star to Earth with at least one confirmed planet is alpha Centauri B which is 4.2 light years away. This particular planet, alpha Centauri B, orbits very close to its star and is almost certainly far too hot to support life; its estimated surface temperature is over 1,000° C (1,800° F).

There are several stars with planets within about 20 light years of Earth which are potential candidates for life (or at least, our kind of life) as shown in Table E. The data are taken from the Habitable Exoplanets Catalogue and also from the NASA Exoplanets Archive.

What makes a planet potentially habitable? Many factors need to be considered when deciding whether a particular planet is a suitable candidate for alien life. Certainly, the planet must be rocky, that is, be solid, unlike the gas giant planets such as Jupiter and Saturn and many others that have been found in other solar systems which have no solid surface at all, being huge balls of gas.

Current thinking is also that the planet must be located at an appropriate distance from its star and hence have a suitable surface temperature so that liquid water can exist. This is known as the habitable zone. However, the realisation that planets and especially moons of gas giants much further away from their Sun may have liquid water under their surfaces due to tidal and gravitational forces has extended the habitable zone considerably.

Table E. Distances of Exoplanets

PLANETS	DISTANCE (light years)
Kapteyn b	13
Wolf 1061c	14
Gliese 876 b,c,d	15
82 Eridani b,c,d	20
Gliese 581 c,d,e	20

In addition, the radiation from the star must be such that the complex molecules needed for life to function are not broken down, for example by strong ultraviolet (UV) light, and the atmosphere must also be appropriate and non-toxic. Venus, for example, a rocky planet close to the inner edge of the Sun's habitable zone, has an atmosphere of carbon dioxide and sulphuric acid with a pressure 90 times of that found on Earth.

Also, the parent star must be of a type that has a long lifetime (of the order of 10 billion years) so that there is sufficient time for life to develop. The lifetime of a star is inversely related to its size (strictly speaking its mass) and can range from just a few million years for truly massive stars to billions of years for really small ones.

We can return now to the original question—'Where are they from?' Twenty light years or so might not sound like much when many astronomical objects are millions and even billions of light years distant, and in cosmological terms it is indeed almost next door. Nevertheless, it is still a huge distance when contemplating a journey, amounting to about 100 trillion miles.

The record for the fastest ever spacecraft is held by

NASA's Juno probe as it entered into orbit around Jupiter in 2016, being accelerated by the giant planet's powerful gravity to a speed of 165,000 mph (265,000 kph). Even at this speed, it would take around 20,000 years to reach the nearest star, Alpha Centauri, at just over four light years away. As mentioned above, this star's planet is an unlikely candidate for alien life so we must look a little further afield at some of the other planets in Table E. (The Parker Solar Probe, launched in August 2018 and designed to orbit and study the Sun, will reach a speed of 430,000 mph (692,000 kph) in 2024 when it will be at its closest to the Sun).

This would imply a journey of up to 20 light years which, with current technology, would take as much as 100,000 years for a one-way journey. We might assume that alien technology would be far advanced compared to ours but even increasing our best speed by a factor of 1,000, taking it to 165 million mph (a quarter of the speed of light), we still have a journey time of around 100 years.

Naturally, we cannot say that this would be impossible for an advanced civilisation but that does seem to be about the shortest journey that a visiting alien would have had to undertake.

2. WHY IS THERE NO PHYSICAL EVIDENCE?

Although some pilotless space probes have been deliberately destroyed on arrival, such as the Galileo probe sent into Jupiter's dense atmosphere at over 16,000 kph (10,000 mph) after its 14-year mission was complete, most end up on the surface of their planet or moon, or at least in orbit around it. Probes with no plans to land anywhere, that is, the Pioneers and the Voyagers have now left the Solar System and are entering interstellar space. They carry plaques as shown in

Mysteries of the Universe

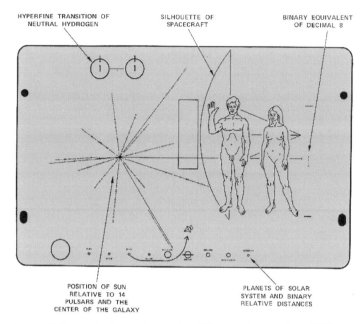

Image 58. Plaque attached to Pioneer 10 and Pioneer 11 spacecraft. ©NASA.

the event that they are ever intercepted by an alien civilisation at some future time.

The message is intended to communicate the location of the human race, the appearance of an adult male and female of our species, and the approximate era when the probe was launched.

More detailed explanations of the meaning of the various inscriptions can be found online. Wikipedia also has a good article entitled Pioneer Plaque.

To me, this is the biggest problem. If advanced alien civilisations do exist, then it would be supremely arrogant to assume that they were unable to solve the problem of travelling the required huge interstellar distances to reach Earth.

However, in my opinion, it is just inconceivable that

visitors from another star system would come all the way here and not leave some indestructible evidence of their arrival, just as our own first travellers did upon their arrival on another astronomical body (Image 59).

So, what are we to make of the numerous photographs, drawings and verbal reports that purport to show UFOs? In the absence of any credible supporting evidence, it is reasonable to assume that the observers were either genuinely mistaken in what they saw, due to atmospheric or other interference, or that they did see what they reported but it was a local military or scientific device (weather balloon, prototype aircraft, satellite, etc.) or that they have deliberately fabricated their 'evidence' for their own purposes.

Finally, we should mention an interesting discovery made in October 2017 followed by another one in August 2019.

Image 59. *Plaque on the landing gear of the Eagle spacecraft left on the Moon by the Apollo 11 astronauts in 1969. ©NASA.*

INTERSTELLAR OBJECTS—OUMUAMUA AND BORISOV

Oumuamua (Image 60), officially known as asteroid 1I/2017 U1, was discovered on 19 October 2017 by the Pan-STARRS 1 telescope in Hawaii as part of NASA's Near Earth Object Observations programme. It became international news because this was the first confirmed object from another solar system to visit ours. Its unusual shape, about 250 metres long and only about 40 metres wide, together with its unique origin, led many people to surmise that it might be an alien craft.

This idea was taken seriously by many astronomers, so much so that significant telescope time was taken up to see whether the object contained any technology. No evidence for this was found. Its speed, quoted by NASA as 196,000 mph (over 300,000 kph), and its trajectory (Image 61) ruled out its origin from within our Solar System.

Then, on 30 August 2019, Gennady Borisov, a Ukrainian amateur astronomer, discovered another object that is now thought to be a comet of interstellar origin. Like Oumuamua,

Image 60. Artist's impression of Oumuamua. ©NASA.

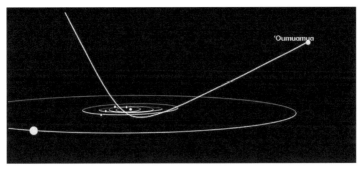

***Image 61**. Trajectory of Oumuamua through the Solar System. ©NASA.*

its trajectory rules out its origin from our own Solar System, and it is now being intensively studied. Comet C/2019Q4 Borisov is especially interesting because it has been found on its journey *into* the Solar System unlike Oumuamua which was only found on its way *out* of the Solar System. This means that while it was only possible to make detailed studies of Oumuamua for about one week, Comet Borisov will be amenable to study for over one year.

It is likely that further objects from other solar systems will be detected due to improvements in both terrestrial and space telescopes. Key indicators will be the speed and trajectory of such objects.

II. ALIEN VISITATIONS

The idea that the Earth has been visited by alien creatures in the distant past who then influenced human development and technology is an intriguing one that was first developed in the late 19th century by various science fiction writers. Its best-known proponent is the Swiss writer Erich Von Daniken who published his book *Chariots of the Gods?* (originally in German) in 1968.

According to his website (www.evdaniken.com), Von Daniken has sold over 60 million books in over 30 different languages. This concept is also the basis for Stanley Kubrick and Arthur C. Clarke's film *2001: A Space Odyssey*.

The 'ancient astronaut' idea is therefore a popular concept and it has spawned many publications from many authors. The primary argument in favour of this thesis is the notion that Earth is littered with numerous artefacts that represent a higher level of technological expertise than what existed at the time they were created. Supporting evidence is supposedly given in prehistoric drawings, some of which are said to resemble modern astronauts and space vehicles.

We can examine a few of the examples given in *Chariots of the Gods?* Sacsayhuaman is a large walled complex near the city of Cusco in Peru. It was built by a pre-Inca people in about 1100, and is remarkable for its walls which consist of

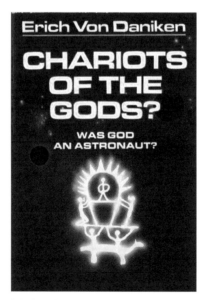

Image 62. *Cover of the first edition of* Chariots of the Gods? *by Erich von Daniken.*

Image 63. Part of the complex wall in Sacsayhuaman, Peru. ©Author.

numerous large stones that fit together so perfectly that in some cases it is not even possible to insert a sheet of paper between them (Image 63).

Von Daniken refers to Sacsayhuaman in his book. He states "Just look at the incredible accuracy of the jointing. How could primitive people handle these huge blocks?"

Similar arguments are used for the many other artefacts around the world.

Look at the beautiful death mask of Tutankhamun created out of solid gold, coloured glass and semi-precious stones by ancient Egyptian craftsmen 3,400 years ago (Image 64).

This and all the other artefacts recovered from ancient Egyptian tombs were created by local craftsmen who had developed their skills over many years. Or do you think that an alien stood by to oversee the work?

These are indeed all marvellous structures but that does not mean that they needed extra-terrestrial assistance in their construction.

***Image 64**. Death mask of Tutankhamun, c.1330 BCE at the Cairo Museum. ©Author.*

It is a big mistake, and an insult, to assume that ancient civilisations were stupid. What they lacked in modern tools and technology they made up for in an enormous labour force, no minimum wage or late penalty clauses and an absence of health and safety legislation. They could employ many thousands of workers on long shifts—if someone was injured or killed, it did not matter. In the end, the job got done.

They had architects and craftsmen who learnt their trade by experimentation. Mistakes did happen though. If a doorway collapsed, it would be cleared and rebuilt in a slightly different way. If a window was created in the wrong place, as in one of the structures in Machu Pichu, Peru, it would be blocked up (Image 65).

Image 65. Blocked-up error (white line) in Machu Pichu, Peru. ©Author.

Another well-known builder's error became manifest in 1178, five years after the construction of the Tower of Pisa had begun. A flawed original design of a completely inadequate

Image 66. Leaning Tower of Pisa. ©Author.

foundation meant that the 55m-high (180-ft) tower, weighing about 15,000 tons, would never remain stable and it started to lean when construction reached the third floor.

So not all ancient monuments are perfect.

Prehistoric art features prominently in 'ancient astronaut' theories and has been used as evidence of first-hand encounters with alien space travellers due to the apparent resemblance of some of the drawings with floating astronauts. Here is a typical example mentioned in *Chariots of the Gods?*

The cave drawing (Image 67) shows two human figures with headgear that does indeed resemble a space helmet with protrusions that could be interpreted as antennae. But that is the problem—interpretation. All manner of masks and headgear were created by ancient civilisations in their homage to the gods and this could be just one more.

It is tempting to compare these rock drawings with photographs of floating astronauts (Image 68), and yes, the comparison is impressive (especially when the angles are adjusted to match)! But there is another much more intriguing problem. Floating astronauts do just that—float. And at an altitude of several hundred miles above the Earth's

Images 67 & 68. *Rock carving, Val Camonica, Italy, c. 8,000 BCE; floating US astronaut.* ©Wikipedia.

surface at a speed of about 18,000 mph. How could our ancestors have seen them?

Look at these two photographs. In 1976, the Viking 1 Mars spacecraft was searching for possible landing sites for future missions when it photographed what became known as the Face on Mars (Image 69, left). The image did indeed resemble a human face and when released by NASA fuelled a great deal of speculation that it had been artificially created by a Martian civilisation.

In view of the intense public interest, NASA programmed its Mars Orbiter camera to take higher resolution photographs of the object in 1998 (Image 69, right).

It turned out that the 'face' was nothing more than a rock formation that, under suitable lighting conditions and low-resolution imaging, had a passing resemblance to a human face.

There are many other features on planetary mission photographs that have been put forward as evidence of extra-terrestrial life but all have turned out to be rock formations

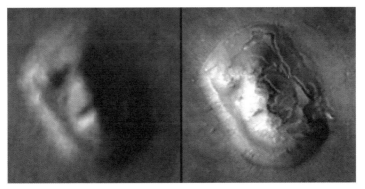

Image 69. 'Face' on Mars, 1976 (left), and 1998 (right). ©Malin Space Science Systems/NASA. ©University of Wisconsin and author.

that, under certain lighting conditions, seem to resemble a human or animal form.

Other artefacts that have been proposed as evidence of past alien visitations include ancient maps. One such item has been prominent in this regard. It is known as the Piri Reis map.

This map was drawn on a gazelle skin by Piri Reis, a Turkish admiral and map-maker, in 1513. A fragment was discovered in the Topkapi Palace in Istanbul, Turkey, in 1929. Marginal notes on the map describe how it was

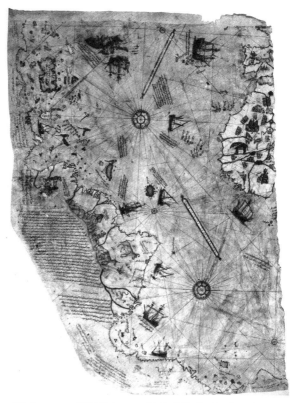

Image 70. *Fragment of Piri Reis map, 1543. ©Topkapi Palace Library, Turkey.*

compiled from a variety of earlier source maps as well as from discoveries made on contemporary voyages by explorers and ships blown off course.

Some proponents of the 'ancient astronaut' theory have suggested that the Piri Reis map contains information that could not have been known by humans in 1513, notably the existence of Antarctica.

The map is not easy to interpret. It is easier to look at the sketch maps (Image 71, left and right) where it becomes clear that although the upper part of South America is quite accurately mapped by Piri Reis, the coastline suddenly diverges eastwards at about the location of Rio de Janeiro.

To interpret this rogue coastline as representing the continent of Antarctica is fanciful. It is in the wrong place, it is the wrong shape, and it is shown as continuous with

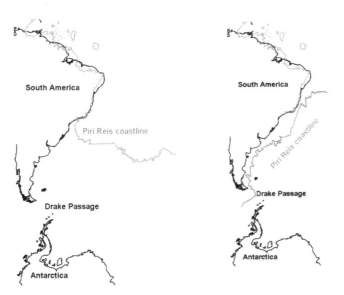

Image 71. Sketch maps of South America and Antarctica with part of the Piri Reis map superimposed in grey. ©University of Wisconsin and author.

South America whereas in reality it is separated by the 600 mile (1,000 km) wide Drake Passage.

The left-hand sketch shows that the first part of the map accurately matches the northern part of the coastline before diverging eastwards. The right-hand sketch shows this divergent coastline bent downwards (by me) to indicate that it also matches the actual coastline accurately. So why did Piri Reis draw the coastline inaccurately?

It is possible that the divergent coastline actually does represent the rest of the South American coastline down to Cape Horn but bent just to fit onto the available space on the gazelle skin. Cartographers sometimes employ this device to make maps fit into available spaces.

A more recent example of this is seen in the following London Underground map from 1908 (actually the first one to show all the lines together in a single map).

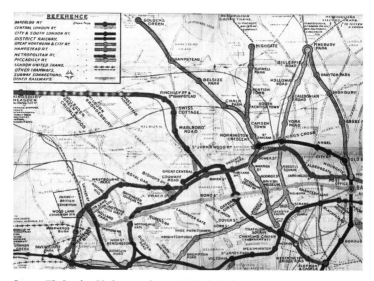

Image 72. London Underground map, 1908. ©Author.

These early London Underground maps showed the tube lines superimposed on a street plan and therefore gave a true representation of the geographical locations of the stations. This design lasted until 1933 when the map was radically redrawn by Harry Beck in a diagrammatic style, which was easier to read but no longer gave a true geographical representation of locations and distances.

The designer of the 1908 map faced a problem as to where he or she could locate the reference chart showing which colours represented which lines. It was eventually placed in the top left corner which meant that the western section of the Metropolitan Line, indicated here by white dots, had to be pushed downwards to create the necessary space.

So even though the map purports to be geographically accurate, the cartographer had to adjust geography to fit in the reference chart. Perhaps Piri Reis had a similar problem of space—his gazelle skin just wasn't big enough.

It should also be mentioned that there had been a belief in a large southern continent since the Ptolemaic times of the first century, and depictions of large southern land masses were common in 16th and 17th century maps. Definitive evidence was eventually obtained by the first sighting of the continent in 1820. So even if Piri Reis's map had shown an Antarctic continent, which it does not, it would not have been something new. See www.uwgb.edu for an excellent discussion of the map.

Ancient statues of strange beings are another class of item cited by Von Daniken and others as evidence of extra-terrestrial visitations, the idea being that these carvings were made in the image of the alien visitors. There are numerous such statues around the world differing in design, size and date. Here are just two.

Mysteries of the Universe

Image 73. Dogu statue, Tokyo National Museum, Japan, c. 1,000 BCE. ©Tokyo National Museum, Japan.

Image 74. Kalasasaya Temple, Tiahuanaco, Bolivia, c. 500 BCE. ©Author.

Image 75. Mount Rushmore Memorial, South Dakota, USA (1927–41). ©Author.

It is true that we have no knowledge about where these ancient sculptors got their inspiration from but we should not assume that they were devoid of imagination. Look at some of our own cultural artefacts, such as the Mount Rushmore Memorial in South Dakota, USA, and the statue of Eros in Piccadilly Circus, London (Images 75 and 76).

What might our descendants think of these 60-ft giant heads should they be rediscovered in a few thousand years' time? We would surely laugh if we thought that they would believe them to be likenesses of 20th-century giant alien astronauts.

Or what about the flying archer statue, Eros, in Piccadilly Circus, London? Flying aliens?

There is no credible evidence for alien visitations to Earth in ancient (or modern) times.

Image 76. Eros, Piccadilly Circus, London. ©*Author.*

III. ALIEN ENCOUNTERS

Seeing what looks like a flying saucer is one thing but perhaps the most fantastic of all claims is that of actually meeting an alien. Such people are known as 'contactees'. There have been thousands of such claims, many of which involve being abducted by aliens rather than just meeting them. Those who are abducted are known as 'abductees'. Such reports date mostly from the 1950s onwards.

George Adamski

A well-known contactee was George Adamski (Image 77), a Polish-American who published numerous UFO photographs, articles and books about his experiences. In 1952, Adamski claimed to have had an encounter in the

Californian desert with an alien from Venus who took him on a ride in their scout ship and then in their mother ship.

Unfortunately, Adamski's camera was unable to work properly inside either craft, the failure being blamed on high magnetic fields. Also, the Venusians would not allow him to take their photographs during their meeting in the desert. That of course was a great shame since his camera would presumably have worked perfectly well on the ground in California. We do however have the benefit of a drawing of this encounter (Image 78).

The sketch of Adamski greeting his Venusian visitor in the desert in California is an illustration from the chapter 'The Story of a UFO Contactee' of the book *How to Make the Most of a Flying Saucer Experience* by Professor Steve Solomon.

One of Adamski's most famous and widely-publicised photographs is of the Venusian scout ship, as shown in Image 79.

Image 77. *George Adamski (1891–1965). ©GAF/International Adamski Foundation, California, USA, with permission.*

Mysteries of the Universe 111

Image 78. *Sketch of Adamski greeting his Venusian visitor in the Californian desert. ©Top Hat Press 1998, with permission.*

Image 79. *Venusian scout ship photographed by George Adamski. ©GAF/International Adamski Foundation, California, USA, with permission.*

The space age had not started in 1952 so detailed knowledge about surface conditions on the planets was not available. Later information from space probes sent to Venus showed that the planet had a surface temperature of nearly 500°C (900°F) which was easily hot enough to melt lead, an atmospheric pressure of over 90 times that on Earth (equivalent to a dive to over 900m or 3,000ft which would require a full body hard diving suit), and an atmosphere containing no oxygen.

The consensus opinion now is that no life could survive on the planet. So, how the remarkably human-like Venusians managed to exist under these conditions seems hard to explain.

Adamski also recounted meetings with beings from Mars and Saturn. Saturn is composed of gas and has no solid surface. It may be theoretically possible for some bizarre life forms to exist deep within the gas but this seems an unlikely habitat for a humanoid creature as described by Adamski.

Much like Erich von Daniken in the 1960s, as discussed earlier in this Question, George Adamski must be given credit for his work which created a huge international following of UFO enthusiasts. He gave lectures around the world and published a number of successful books about his exploits.

His best known is probably his first—*Flying Saucers Have Landed*—written with his friend Desmond Leslie and published in 1953 (Image 80).

It provides a detailed account of his meetings with the aliens and includes several photographs and drawings. The book is out of print but an electronic copy is available for download at www.universe-people.com. It can also be bought second-hand from Amazon. The excellent cover art shows the scout ship hovering over the Californian desert with the mother ship waiting in the background.

Image 80. Book cover—Flying Saucers Have Landed.
©*Desmond Leslie & George Adamski.*

Adamski's fan base shrank considerably when, in 1959, he announced that he was going to a conference on Saturn. Even for him, that was probably a giant leap too far.

BETTY AND BARNEY HILL

One of the best-known alien abduction reports is that of Betty (1919–2004) and Barney Hill (1922–69). On 19 September 1961, they were driving home through the White Mountain area of New Hampshire. They saw a white light in the sky and stopped the car. Barney got out and went to study the white light more closely with binoculars. He became convinced that he had seen a UFO.

Both described it as a pancake- or banana-shaped object. Barney stated that he could see beings behind the craft's windows. Soon after, they returned to the car and drove home. The couple reported their experience to the nearby Pease Air Force Base and also to a national UFO group.

Both the Hills began suffering from a variety of ailments. Betty had a series of nightmares involving beings with catlike eyes and Barney had a painful back. They could also not account for two hours of their time during their journey home.

Inside the craft, the aliens carried out various medical experiments and took some tissue samples. Finally, the couple were taken on a tour of the spacecraft. Betty asked whether she could take a book as evidence of the encounter but unfortunately this was denied. She asked where they came from and was shown a star map which she later recalled under hypnosis (see Image 82).

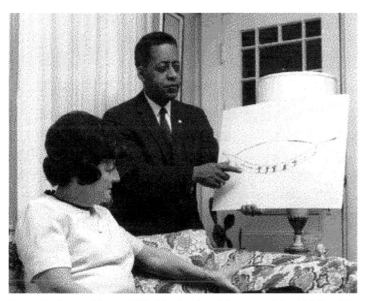

Image 81. *Betty and Barney Hill with a drawing of the spacecraft in which they claim to have been abducted. ©Betty and Barney Hill.*

In 1964, three years after the event, they were referred to a psychiatrist who recommended a session of hypnotic regressions. Under hypnosis, the Hills revealed that the UFO had landed close to the car and they were somehow partially sedated. They were then forcibly taken aboard the spacecraft by beings with large catlike eyes and white skin.

The Hill case became worldwide news and spawned a book, *The Interrupted Journey* by John Fuller, as well as a 1976 TV movie, *The UFO Incident*.

It would be easy to dismiss this case as a fantasy brought on perhaps by tiredness and the sighting of a military aircraft from the nearby Pease Air Force Base. Easy, except for one thing. Under hypnosis, Betty Hill recreated the star map she said she had been shown by the aliens. Two years later, in 1966, Marjorie Fish (1932–2013), a school teacher and amateur astronomer, wondered whether the Hill star map bore any resemblance to an actual star pattern. Her interpretation is shown in Image 82, right, which should be compared with Betty Hill's drawing.

The similarities are uncanny and led many people to believe that story was true and that the aliens encountered by the Hills came from the binary star system Zeta Reticuli, which is about 39 light years from Earth.

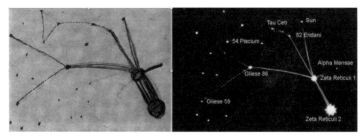

Image 82. *Betty Hill's star map recalled under hypnosis, 1964 (left); Marjorie Fish's interpretation, 1966 (right). ©Author.*

It is particularly impressive when one considers that some of the stars in the drawing were unknown until the Gliese star catalogue of 1969 was published. It should also be mentioned that no planets have been detected so far in the Zeta Reticuli system, although of course that does not mean they do not exist.

So, what is the explanation? How did Betty Hill manage to draw an apparently accurate map of a distant star system not only two years after she said she saw the original diagram in the aliens' spaceship, but also five years before some of the stars had even been catalogued?

As always, we have more than one explanation and it is prudent to heed William of Ockham's advice and go for the simplest explanation that makes the fewest assumptions.

An excellent analysis was given in the December 1974 issue of *Astronomy* magazine in an article entitled "The Zeta Reticuli Incident" by Terence Dickinson. This included a

Image 83. *Night sky.* ©*Author*

section by Carl Sagan and Steven Soter in which they gave a detailed critique of Fish's interpretation. Their main point was that both the Hill drawing and the Fish interpretation looked similar because of the lines connecting selected stars. Remove the lines and replace some of the stars that had been left out of the published drawings and the similarity disappears. If one re-joins different stars with different lines the similarity also disappears.

Image 83 is a photograph of the night sky taken in the Canary Islands. Well over a hundred stars are visible and it is clearly possible to join selected ones together with lines to make different shapes and patterns, which is how the ancients devised names for the constellations.

With so many stars to choose from it would be easy to create similar patterns from different star fields by just selecting those stars that made the pattern you want.

The basic idea is illustrated Images 84 and 85.

Image 84 shows two star fields but to keep it simple, there are just 15 stars (white dots) in each half. Parts A and B are meant to represent two different star fields, that is, different parts of the sky with stars in completely different positions.

Image 85 shows the same star fields as Image 84, but some of the stars have been connected with lines.

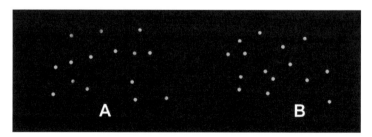

Image 84. *Two star fields.* ©*Author.*

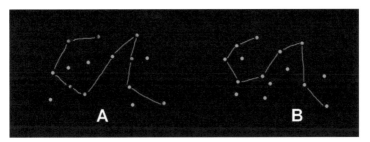

Image 85. *As image 84 but with selected stars joined up. ©Author.*

Even though A and B have stars in completely different positions, it is still possible to create similar patterns in both by careful selection of which stars to connect.

This is possibly why the Fish interpretation of the Hill map was such an impressive likeness. With a large number of stars to select from, it would be a simple matter to join just those that repeated Betty Hill's original pattern.

To make the comparison even more persuasive, it would have also been easy to remove many of the superfluous stars that only appeared in one of the illustrations. The result would be a good reproduction of the Hill pattern and with many of the unconnected extra stars in similar positions.

There have been many more reports of human–alien encounters but these will suffice as a fair sample.

With photographs easy to fake, and with no photographs at all of Adamski's alien friends, and with no alien artefacts to examine, one might be tempted to dismiss his accounts of the encounters because of lack of credible evidence. This is a science book so credible evidence is key.

As to the Hill encounters, which is the more reasonable explanation? If you want to believe in UFOs and aliens, then you would choose the Hills' version but if you prefer to make your decision based on credible evidence and what seems

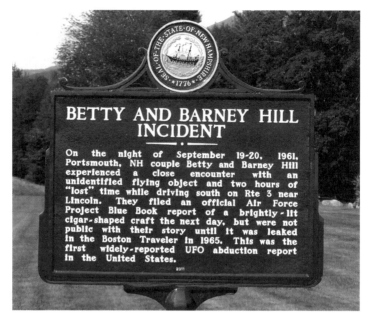

Image 86. *Signpost commemorating the Betty and Barney Hill incident. Source: Wikipedia.*

reasonable, then you might choose Professor Sagan's version. It should also be mentioned that Marjorie Fish subsequently retracted her interpretation after new astronomical evidence became available.

Most of Betty Hill's notes, tapes and other items are in the permanent collection of the University of New Hampshire. In July 2011, the state Division of Historical Resources marked the site of the alleged craft's first approach with a signpost.

IV. CONCLUSIONS

If there was good evidence for the existence of UFOs, or for visitations by aliens, or for alien abductions, then this question would already have been answered. But since there is no good evidence for any of these happenings, we can end the topic of alien life with the question as to whether it exists at all.

Before doing so however, it might be interesting to give a very brief overview of aliens in an historical context. Lucian of Samosata (c. 120–180 CE) is widely regarded as the author of the first science fiction story. His *True History* (still available online and on Amazon) tells the story of a group of adventurers who are blown off course and land on the Moon where they become embroiled in a war between the lunar people and those living on the Sun.

Ancient explorers encountering new territories on Earth would not have been surprised to find them inhabited. It was therefore quite reasonable for people to think that the Sun and the Moon, the only astronomical objects that could be seen to have a surface before the invention of the telescope in 1610, were also inhabited. (Strictly speaking, a Dutch-designed instrument was made in 1608 by Hans Lippershey but this was not used for astronomical observations. Galileo modified and improved on Lippershey's design and produced the first astronomical telescope two years later).

Sir William Herschel, the discoverer of Uranus in 1781 and Astronomer Royal to King George III, also believed that the Moon and the interior of the Sun were populated. Later, in 1877, the Italian astronomer Giovanni Schiaparelli was mapping the surface features of Mars and saw what he called *canali* (Italian for 'channels') on the surface.

This became mistranslated into English as canals, implying construction by intelligent beings. The then widely reported completion of the Suez Canal (in 1869) meant that the mistranslation was interpreted as the discovery of large artificially created structures on Mars.

American astronomer Percival Lowell was also convinced

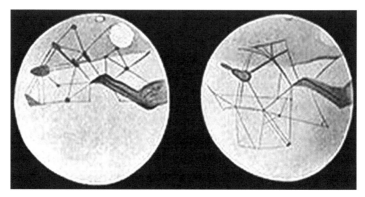

Image 87. Lowell's sketch maps of Martian 'canals', c. 1895. ©Wikipedia.

that he had seen canals on Mars and produced detailed drawings (Image 87).

Orson Welles's famous radio broadcast of H. G. Wells's *The War of the Worlds* was apparently so realistic that many listeners actually believed that this was a live news bulletin

Image 88. Newspaper headline about Welles's broadcast. ©Wikipedia.

and that Earth was being invaded by Martians. (Whilst some listeners did probably believe this to be true, the stories of mass panic seem to have been exaggerated for publicity purposes). Nevertheless, the event does show that there was still a belief in Martian life in 1938.

Returning now to the question of whether alien life exists at all, we have three possibilities:

1 Yes, there is life elsewhere in the universe
2 No, there is no life elsewhere in the universe
3 There was life elsewhere in the universe but it is now extinct (wiped by war, disease, parent star explosion, astronomical impact, etc.)

The basic argument in favour of extra-terrestrial life is simple. With billions of stars and billions of galaxies, there must be life somewhere else as well as on Earth. Many scientists and lay people subscribe to this view. If that was true, how would we communicate with each other? The obvious answer is by radio transmission, probably at the frequency at which hydrogen atoms emit radio waves (1420 MHz at a wavelength of 21cm). Why? Because hydrogen is the simplest element and is therefore a likely choice.

Communication with Extraterrestrial Civilisations

The first primitive radio telescopes, devices which could receive radio signals from space, were produced in the 1930s. The rapid development of radar during World War II then stimulated the formation of a new field of research—radio astronomy.

Professor Frank Drake, an American astronomer and astrophysicist, had a deep interest in the subject of extra-terrestrial life and how we might be able to communicate with it, and founded the organization known as Search for Extra Terrestrial Intelligence (SETI). This was set up in Green Bank, West Virginia, USA, in 1961 and its radio telescope began to listen for transmissions from space in the hope of detecting messages from alien civilisations. SETI's website (www.seti.org) has a number of interesting articles and a 'Frequently Asked Questions' section, and is well worth visiting.

Professor Drake devised an equation, known as the Drake equation, to try and work out how many such civilisations existed in our galaxy with which we might be able to communicate.

Drake's pioneering idea was to consider those factors which were likely to be important in the development of intelligent life, and to put these together in a solvable equation. 'Intelligent life' meant life with which we could, in theory, communicate. This then requires an alien civilisation

Image 89. Professor Frank Drake (b. 1930) with his equation. ©Frank Drake.

that has developed the technology to use radio waves for this purpose.

Here is what Professor Drake came up with in 1961.

$$N = R \times f_p \times n_e \times f_l \times f_i \times f_c \times L$$
The Drake Equation

This is what the various symbols mean:

N is the number of civilisations in our galaxy with which we could communicate. This is the number we are trying to calculate.

R is the number of new stars of the same type as our Sun that are formed annually in our galaxy.

f_p is the percentage of Sun-like stars that have planets.

n_e is the average number of planets, for every Sun-like star that has planets, that could support life.

f_l is the percentage of planets out of those that could support life that actually have developed life.

f_i is the percentage of planets with life that have developed intelligent life.

f_c is the percentage of intelligent life forms that use communication technologies compatible with ours, that is, electromagnetic transmissions such as radio waves.

L is the average amount of time, in years, that such a civilisation exists before becoming extinct or being destroyed.

This was an innovative concept and was the first serious attempt to quantify the likelihood of making contact with extra-terrestrial civilisations. There is, however, a really big problem with this equation that, in many people's views, unfortunately renders it valueless.

Although it is possible to make educated guesses based on actual research data for some of the values in the equation, most are utterly unknown and unknowable.

There is absolutely no information at all for the last four components of the equation.

It is a pure guess based on your own gut feeling.

You could put in any value at all for these components and they would be just as valid as any other. Your guess is as good as anyone else's.

The problem is exacerbated because the factors have to be multiplied together. Let us assume that the current estimates for the first three components are reasonably accurate. Let us also be very generous and assume that the values for the last four components are in error by a factor of just ten. That means that the value of N is in error by a factor of 10 x 10 x 10 x 10 = 10,000. However, if the guessed values for the last four components are in error by a factor of 100, then the final result is wrong by a factor of 100 million!

The trouble is that with just one example of life having formed (on Earth), we cannot make any assumptions about how likely it was. If we should find life somewhere else, Mars maybe, or one of the moons of Jupiter, then the whole situation changes and we could justifiably say that life is likely to be quite common under the right conditions.

Drake's original SETI conference in 1961 gave some 'educated guesses' for the various values of the equation. Using the lowest values given, the value of N, the number of intelligent civilisations in the Milky Way galaxy, calculated at about 20. However, using the highest values given, the equation predicted nearly 40 million advanced civilisations. Even over 50 years later, it is still impossible to solve the Drake equation since the values of most of the components are utterly unknown. According to an article in *New Scientist* (31 August 2019), the latest estimates range from 1 (that is, just us) to 4 billion.

So, until we know more about the properties of exoplanets, and until we find some evidence of extra-terrestrial life, the Drake equation remains unsolvable. Having said that, the equation was the first serious attempt to quantify the likelihood of contacting an alien civilisation and Professor Drake has rightly been given great credit for his pioneering work.

There is, however, an intriguing alternative possibility that might produce results far quicker than waiting many decades for the necessary evidence from other moons and planets to fit into the Drake equation.

The basic biochemistry of all life on Earth, whether viruses, bacteria, plants, trees, fish or mammals, is very similar.

It is based on nucleic acids, RNA and DNA, to pass genetic information on to offspring, and it depends on proteins, fats and carbohydrates to keep itself alive and well. This has led to the currently accepted view that all life on Earth has a single common ancestor from which everything else developed. Put another way, life started on Earth just once.

Question 3 – How did Life Begin? explained the concept of molecules that are either left- or right-handed, like a pair of gloves.

Supporting evidence for a single common ancestor is that all sugars found in living creatures are of the d (right-handed) form, and all proteins are made of amino acids of the l (left-handed) form. If life had started several times, then it would be reasonable to expect some of it to have a different biochemistry, one version of which could be a different chirality, that is, it might be using left-handed sugars or right-handed amino acids. See also pages 39 to 40.

Such life forms would look the same as their other-handed

counterparts but they would be completely incompatible with them. An apple, for example, based on a reverse-handed biochemistry, would be indigestible and therefore completely useless as food for a creature with a biochemistry based on existing handedness.

All life so far found on Earth has d sugars (right-handed) in its carbohydrates and l amino acids (left-handed) in its proteins. If even one example of an organism is found with a different mechanism for passing genetic material, or with all of its proteins constructed of right-handed amino acids, then that would be good evidence that life had arisen on Earth more than once. If it happened more than once here, then it could also have happened elsewhere. We can then make a better guess at some of the components of the Drake equation.

For now, all we know for sure is that life has arisen once, and any results from the Drake equation must be considered invalid.

SETI has been searching for extra-terrestrial life for over 50 years and so far there has been no irrefutable evidence for any radio signals that could be construed as a message. The so-called Wow! signal (shown later in this chapter) was a one-off event and in the absence of any repeat signals has to be considered as a technical rather than an extra-terrestrial event.

The Breakthrough Initiatives

In July 2015, the Russian entrepreneur Yuri Milner, in collaboration with leading astronomers and other scientists, announced his Breakthrough initiatives.

Breakthrough Listen will buy time on the World's largest telescopes and search for radio and laser transmissions that may come from extra-terrestrial civilisations.

Breakthrough Message will study the ethics of sending messages into space. It also launched an open competition with a US$1 million prize pool, to design a digital message that could be transmitted from Earth to an extra-terrestrial civilisation.

Breakthrough Starshot was announced in April 2016 and is a $100-million programme to develop a proof-of-concept light sail spacecraft fleet capable of making the journey to the Alpha Centauri system at 20 percent of the speed of light (135 million mph; 215 million kph), taking about 20 years to get there and about four years to notify Earth of a successful arrival.

Breakthrough Starshot will attempt to aim its cameras at the recently discovered exoplanet Proxima b with the hope of capturing some surface features. The spacecraft fleet would have 1,000 crafts, and each craft would be a very small centimetre-sized device weighing just a few grams. They would be propelled by several ground-based lasers of up to 100 gigawatts.

The concept is illustrated diagrammatically in the following images (from Mr. Milner's press conference).

Further information on the Breakthrough initiatives is available on the internet including a detailed description on Wikipedia.

***Image 90**. Laser array; light sail with spacecraft held in its centre. ©Yuri Milner.*

The Probability of Extraterrestrial Life

As stated at the beginning of this chapter, the usual argument in favour of extra-terrestrial life is straightforward. With billions of stars and billions of galaxies, there must be intelligent life somewhere else as well as on Earth. We cannot work out the probabilities since, as we have seen with the Drake equation, we do not have sufficient information. But we can put it into some sort of context.

We need to make an assumption, which is that the life we are talking about is such that it has to arise on a rocky planet revolving around a star. We might be able to conceive of some bizarre life forms that could exist in space, or inside stars, but as we know absolutely nothing about such theoretical life forms, we are not in a position to consider them.

The problem is that it is impossible to work out the probability of something happening if it has only been observed once and you do not know how many possibilities there are. Let us imagine that you put your hand into a box and draw out raffle ticket number 141. What were the odds of you picking that particular ticket?

It is impossible to say if you do not know how many tickets are in the box. In the same way all we know about life is that it is possible—we have no idea of how likely it was.

Earthbound telescopes, the Hubble space telescope and various spacecraft have made extensive studies of every planet and many moons in our Solar System.

To date there has been no evidence of any life, intelligent or not. The most detailed studies have been on Mars where robot landers have been surveying the landscape, digging up the soil and analysing it since the 1970s. The latest lander is NASA's Curiosity rover which set down in August 2012. Its task is to look for signs of past and perhaps present life.

Some photographs of the Martian surface have shown what could be dried-up rivers and lakes, leading some people to suggest that there may have been liquid water on the surface at some time in the past. The Jovian moon Europa seems to be covered with a thick layer of water ice which has led to speculation that there may be liquid water underneath. There is even some evidence that water may have been present on our own Moon in the past.

Life depends on chemical reactions. Living creatures need to feed, grow, repair themselves, reproduce, and interact with their environment. All this needs chemistry, and chemistry needs to take place in some type of support medium. For all of us on Earth it is water.

The absence of liquid water on other planets and moons makes it unlikely that life ever developed there. Not impossible of course because other substances such as liquid ammonia or hydrogen sulphide could replace water as the medium.

It is easier however to start with what we know, so the presence of water on other planets and moons is an exciting discovery since it at least makes it possible that some sort of life may have developed there.

Earth is a small rocky planet. So are Mercury, Venus and Mars (and Pluto although it is now classified as a dwarf planet).

Space probes have visited all the planets and many of the moons in our Solar System. There has been no evidence of life on any of these planets or moons. They are either too close to the Sun and therefore too hot, or too far away and therefore too cold. As it turns out, the Earth is in just the right position being neither too hot nor too cold for life to survive. Had it been a bit nearer to the Sun or a bit further away, we would not be here now. Life can only exist within a very narrow range of values.

Mars, which depending on the relative positions, is sometimes the closest planet to Earth, has been extensively studied for signs of life. Although nothing has been found so far, the Curiosity lander is currently analysing rock samples for evidence of past and perhaps even present life. The ExoMars mission planned for 2022 will also photograph, drill into and analyse samples taken from below the Martian surface. If anything is or has been living there, it is hoped that Curiosity and/or ExoMars will find it. We will have to wait and see.

The search for extra-terrestrial life now moves on to the possibility of other solar systems, that is, planets orbiting other stars. We have planets orbiting our star, the Sun, but how common are planets around other stars? Again, with just one example, our own star, we could not say. It may be very common or exceedingly rare.

This all changed in 1992 with the discovery of the first exo- (outside our Solar System) planet. NASA's Kepler spacecraft, launched in 2009, is designed to search for exoplanets by measuring the slight loss of starlight as a planet passes between the star and the observer. This is known as a transit and is measured by the Kepler telescope.

These four diagrams illustrate the passage of an orbiting planet around its star. As the planet comes around, it will begin to block out some of the starlight (second from left).

Image 91. Transit of a star by a planet. ©Author.

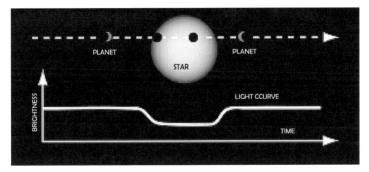

Image 92. *Light curve created by detecting the light loss of a passing planet in front of its star. If Kepler registers three consecutive and consistent light losses then this is taken as confirmation of a planet.* ©*Wikipedia.*

This blocking continues until the planet eventually passes back behind the star.

The loss of light is measured by instruments on Kepler which then creates a light curve as shown in the diagram above.

Not only can new planets be detected by this process, it is also possible to determine the size of the planet, how long it takes to orbit its star, how far it is from its star, what it is made of, and, if it has an atmosphere, what that is made of. There are also various other methods for detecting exoplanets such as looking for a wobble in the motion of the parent star and, in a small number of cases, direct visualisation of the planet itself. (See 'Where are they from?' on page 89 for more on exoplanets).

The known exoplanets vary in size from Earth-size up to much larger ones, some even exceeding the size of our largest planet, Jupiter. Most interestingly though, several of the new planets were located in the habitable zone around their star, a zone where liquid water—and perhaps life—could exist. Some of these are similar in size to the Earth.

So, we can safely say that planets are common, and that some stars have Earth-sized planets circling in the habitable zone. It does seem likely then that our galaxy, and presumably all other galaxies, are full of planets, and that some of these will be small and rocky and at the right distance from their star for liquid water to exist. There would be many other conditions that also need to be right, a recently suggested one being the presence of the correct amount and type of UV light from the host star. The UV light is thought to be essential to kick-start the organic chemical reactions necessary for primitive life.

We therefore have another zone—the abiogenesis zone—planets where there is the right amount of UV light to activate organic chemistry. Abiogenesis means the emergence of life from non-life.

Given the right planet at the right distance with the right ingredients, how likely is it that life will arise?

This is the big question that we struggle to answer. Just because a planet or moon is in the right place for life to arise does not mean that it will.

It has happened once to be sure, but we have no clues whatsoever as to how likely it was. All we know is that it took about 1 billion years. (It is f_l in the Drake equation).

As far as we know, our potential life-containing exoplanet needs to satisfy the following minimum conditions.

i) have a stable host star with a long enough lifetime for life to begin and evolve;

ii) be located within the habitable zone of its star where liquid water can exist, or if further out from its star, to have subterranean liquid water caused by tidal forces from the star and possibly from its planet;

iii) be in the abiogenesis zone so that there is the correct amount and type of UV light to initiate the organic chemical reactions necessary for primitive life;
iv) contain the basic chemicals and nutrients necessary for life;
v) have a moon that is large in comparison to its own size so that its gravitational influence helps to stabilise the planet's spin, resulting in a relatively stable climate over long periods of time. Earth's moon is the largest planetary satellite in the Solar System compared to its host planet. Its diameter is 28 percent of Earth's. (Pluto, now classified as a dwarf planet, has a moon, Charon, that is nearly half its size).

As stated in v), a large moon is thought to be critical. Of the close to 200 presently-known moons in the Solar System, ours is the second largest in comparison to its host planet. This comparative size, rather than just the overall size, is important because of the effect of the Moon's gravity on its planet. Ganymede, which is the largest moon in the Solar System, only has less than 0.01 percent of the mass of its host planet Jupiter and will have insignificant effects on the orbit of Jupiter. Our Moon, however, has over 1.2 percent of the mass of the Earth and its gravity therefore has a significant stabilizing effect on Earth's orbit, ensuring a degree of climate stability and enabling life to develop.

The relative sizes of moon and planet are suitably demonstrated in Images 93 and 94.

Moons are thought to arise in one of two ways. Small moons were probably wandering space rocks captured by the gravitational pull of the planet whereas larger ones may have formed after a collision between a wandering object and a planet.

Image 93. *Jupiter and Ganymede, ©NASA.*

Image 94. *Earth and the Moon, ©NASA.*

It is thought that our Moon was formed in this way, known as the Giant Impact Hypothesis, and illustrated in Image 95.

In this scenario, the Earth was in collision with a massive object the size of Mars and the Moon was formed from the resulting debris. This collision is estimated as having occurred about 4.5 billion years ago. It would of course have been a completely random event but, as outlined above, a very fortunate one since without it, life may never have developed.

Finally, mention must also be made of Earth's magnetic field, caused by the motion of its metallic molten core. The field extends beyond the atmosphere in what is known as

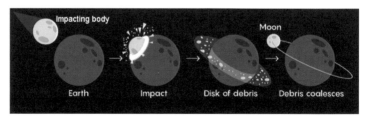

Image 95. *Giant Impact Hypothesis for the formation of the Moon. ©Wikimedia Commons.*

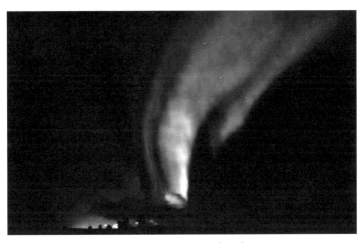

Image 96. Aurora Borealis in Reykjavik, Iceland. ©Author.

the magnetosphere. This has an irregular shape depending on the position of the Sun but is many tens of thousands of miles deep. It protects the Earth from bombardment by numerous particles and cosmic rays which would otherwise by extremely hazardous to any life. Not all planets have magnetic fields, Mars and Venus being examples.

The interaction of cosmic rays and charged particles from the Sun with our magnetosphere is responsible for the auroras often seen in northern latitudes.

The Recipe for Life

Life exists on Earth so the conditions for this to have happened must have been correct. Although we do not know precisely what these conditions were, or which of them were critical, we can make some educated guesses as to the most important conditions and ingredients.

We need:

a planet

1. with a solid surface
2. in a solar system with a stable star that has a long lifetime (at least 10 billion years)
3. located where liquid water can (and does) exist
4. with a magnetic field to shield its surface from harmful cosmic rays
5. with a large moon relative to its size to ensure orbital stability
6. with an atmosphere conducive to biochemical processes
7. with sources of carbon, other elements and compounds necessary for complex molecules to be formed

What we cannot reliably estimate is the likelihood of these conditions being met. However, let us try and make some estimations. Current estimates indicate that there are perhaps 100 billion (10^{11}) galaxies in the universe, and that a typical galaxy may contain about 500 billion stars (5×10^{11}). This gives us a figure of 5×10^{22} for the total number of stars in the universe.

Stars come in different types ranging from supergiants to white dwarfs. Giant stars and small dwarf stars may not be life-sustaining (wrong type of radiation; too short a lifespan; and other reasons), and only about five percent of all stars are in the same class (spectral type G) as our Sun. So out of the total number of stars, there may be about 2.5×10^{21} that are similar to our Sun.

How many of these stars will have planets? Let us be generous and say that they all will. But how many will have a small rocky planet located in the habitable and abiogenesis zones that contains the ingredients and atmosphere and large moon needed for life to develop?

To date (2020), NASA's Kepler telescope has discovered over 4,000 exoplanets. However, the area of sky examined by Kepler is about 0.25 percent of the entire sky so it would require 400 Keplers to cover the whole sky, meaning that around 1.6 million exoplanets might have been discoverable. The Transiting Exoplanet Survey Satellite (TESS) has been searching for exoplanets since its launch in April 2018 and covers a far wider field of view than Kepler. It is expected to find more than 20,000 exoplanets during its two-year mission.

As we have seen, apart from being in the habitable zone, a potential life-containing exoplanet must also be in the abiogenesis zone, and the two do not necessarily overlap. Researchers at the Arecibo telescope facility in Puerto Rico looked at the numbers of exoplanets out of the 4,000 plus known ones which were in their star's habitable zone and found 49. Out of these, only eight were also in the abiogenesis zone, and only one, at most, was likely to be a rocky planet rather than a gas giant. Only two planets are known for certain to be in both the habitable and the abiogenesis zones, Earth and Mars. And as far as we know, there is no life on Mars.

So, although there are undoubtedly vast numbers of exoplanets in our, and presumably other, galaxies, ones suitable for life may be very rare indeed. Based on the above data, the most optimistic figure currently would be one in 4,000.

This then gives an upper limit of 2.5×10^{21} divided by 4,000 which equals about 6×10^{18} (6 million million million) potential life-sustaining planets in the known universe. On how many of these planets will life actually begin to evolve?

We know of one, but have no evidence for any others. It may be very common, or it may be very rare, or indeed it may be unique to Earth. At our current state of knowledge, we have no way of knowing, and this is the problem with trying to solve the Drake equation.

As discussed in Question 3, a succession of events, that is, a process had to occur for life to begin on Earth. Such a process, while unlikely to be the same as that which took place on Earth, would also have to occur on an alien planet for life to begin.

Now consider this. Let us replace the process that led to the origin of life on Earth with an imaginary robot. Imagine also that there is such a robot on each of the 6 million million

Image 97. *Coin-tossing robot. ©Author.*

million potential life-sustaining planets in the universe, and that these robots have been programmed to toss a coin and to record the results, either heads or tails. It is a fair coin and a fair toss.

What are the chances of a coin coming up heads 62 times in a row? It is obviously possible but it is very unlikely. Actually, the chances are 1 in 2^{62} which is equal to about 1 in 6 million million million. This is the same as the calculated number of potential life-sustaining planets in the universe (see page 138) which is why the number 62 was chosen.

In other words, somewhere in the universe, on one of the planets circling one of the stars in one of the galaxies, it is likely that one of the robots would have recorded 62 consecutive heads.

Tossing a coin is a very simple process since there are only two outcomes. It is easy to record the results and it is easy to work out the probabilities of different outcomes. In contrast, the origin of life is a very complex process of which we understand very little. It is easy enough to record the result—a living organism—but much harder to work out the probabilities. In fact, it is impossible. We only know of one occurrence and that does not give us enough information to work out how likely it was.

Think of dice. Dice usually have six sides numbered from 1 to 6.

The chance of rolling any particular number is then one in six, just like the chance of a coin coming up heads is one in two. One can however obtain dice with more or less than six sides as shown in Image 98.

It is easy enough to work out the odds of throwing any particular number with any of these dice. But imagine a dice with an unknown number of sides numbered from 1 upwards.

Image 98. *Dice with different numbers of sides.* ©*Wikipedia.*

Let us say we roll it and it comes up 4. What does that tell us about the chances of it coming up 4 a second time? Absolutely nothing.

We do not know how many sides it has so we cannot work out the probabilities. The fact that it came up 4 on its first roll tells us nothing. If it does come up 4 again on a second roll, then we would have some history to work with.

It is the same with the origin of life. All we know is that is has happened once. That only tells us that it is possible but it tells us nothing about how likely it was.

What if it was as likely as a coin coming up heads 62 times in a row? In that case, as we have seen with our imaginary robots, it would probably only have happened once in the entire universe.

It has been said many times that with so many billions of stars and galaxies in the universe, it is almost certain that many of them will have planets with life. We really are not entitled to draw this conclusion since we do not know how likely it is.

Given enough time, life may develop on every suitable planet or it may be a chance event, like the consecutive head tosses, which while possible is so unlikely that it has probably only ever happened once.

It is perhaps worth reiterating the commonly-held view amongst both the lay public and the scientific community that there are likely to be so many potential life-sustaining planets in the universe that it seems inconceivable that ours is the only one harbouring life.

A few pages back there is a calculation that gives a best guess as to the number of potential life-sustaining planets in the known universe. Written out in full, this number is 6,000,000,000,000,000,000. In words, it is six million million million. That is indeed a lot of planets.

Evidence indicates that life started on Earth just once and that it took about 1 billion years to happen. Let us say that we place a telephone order for a book.

We may expect it to arrive a few days later. If it arrived on the same day we would be astounded and if it arrived three months later we would be disappointed.

Based on experience, we have a likely time frame for the book's arrival. We have no such experience or time frame, for the emergence of life and therefore have no way of judging whether 1 billion years is astoundingly fast or disappointingly slow.

Our book might arrive on the same day if the necessary circumstances prevailed. For example, the book could already be available in the despatch department due to a cancelled order and the courier could leave on a route that includes our address. On the other hand, if the book is out of stock then delivery may well take several months or even longer. In other words, the speed of delivery depends on the circumstances.

It is easy enough to compile a list of circumstances that would determine the speed of our book's arrival. On current knowledge, we cannot compile such a list for the emergence of life, at least not a complete one.

Our Sun has a lifetime of about 10 billion years (see Question 9) after which it will turn into a red giant and engulf the inner planets, including Earth. So if 1 billion years turns out to be astoundingly fast for life to appear, and the 'normal' time is nearer ten times that, then it is clear that we have been very lucky to have had such a superfast delivery. Clearly, if the 'normal' delivery time is closer to 10 billion years, the Earth (as well as similar alien planets circling other Sun-type stars) would not have been in existence long enough for life to have formed.

On the other hand, if 1 billion years is disappointingly slow, and the 'normal' time is one-tenth of that, then it is likely that life would have arisen on many other planets circling Sun-type stars because there would have been easily enough time for the process to occur.

Unfortunately, we have no guidelines as to the 'normal' delivery times for the emergence of life and are therefore unable to gauge its likelihood. So even though there may be millions and millions of potentially life-sustaining planets in the universe, it still does not entitle us to make any assumptions at all as to whether any of them harbour life.

Their sheer numbers, while impressive, do not on their own give us sufficient information to make a judgement. The point that I am trying to make here is that there are many things that are so unlikely that, although theoretically possible, have almost certainly never happened. Here is a well-known example.

Dealing a perfect bridge or whist hand (each player receiving 13 cards of a single suit) from a properly shuffled deck

The odds of this happening are about 1 in 2×10^{27}, or 1 in 2,000,000,000,000,000,000,000,000,000 when written out in full. To put this into perspective, if the present world's population of about 7 billion people had all been dealing bridge hands at a rate of one complete deal every 10 seconds since the universe began 13.7 billion years ago, then by now they would have dealt only about 10 percent of all possible hands. So even with that many deals in that amount of time, it is still very unlikely that four perfect hands would have appeared.

You may have read about players each receiving perfect hands and there are indeed several press reports of such incidents. You have to ask yourself whether such staggering odds are likely to have been beaten, not just once but several times, or whether there is another more likely explanation. In my opinion, no set of bridge (or whist or other similar game) players has ever been dealt four perfect hands from a fairly and properly shuffled deck. Honest reports of such events would have been due to either a false shuffle of a pre-arranged deck which, when well executed, would be undetectable, or by swapping the deck for a pre-arranged one. Cutting the deck does not alter the arrangement of the cards and is irrelevant.

Imagine for a moment that mathematical methods for calculating odds are not available, and someone asks whether it is likely that a perfect bridge deal has ever happened. Not knowing the odds, and assuming that there are many millions of players who have each played many thousands of times resulting in many billions of deals, you may be tempted to say, "Yes, with so many deals it must have happened."

Almost certainly you would be wrong. You have assumed that "many billions of deals", an impressively large number, is enough to cater for all possible deals, which we know from the actual odds, is not even nearly true. It is the same with the estimated number of potentially habitable planets in the universe—6 x 10^{18} another impressively large number.

However, since in this case we really do not know the odds (of life forming), we cannot say that there are enough planets for it to have happened again outside the Earth.

At present there is no evidence of extra-terrestrial life. Marconi was awarded a patent for his radio system in 1896, and regular radio broadcasts from the BBC began in 1922. An alien within about 45 light years of Earth who was listening for signals from our Solar System would have begun to hear remnants of these radio transmissions from the late 1960s, or earlier if they were nearer. Any intelligent life within this distance would then have had time to send a response which we would have received by now.

There are several thousand stars within 45 light years of our Solar System, and probably several hundred that are in the same spectral class as our Sun. Table F shows a list of stars within 45 light years of Earth that have confirmed planets.

Radio astronomers have been listening for signals from stars for about 50 years and continue to do so. That is right within the time frame for the planetary systems listed in Table F, but so far, they have not come across anything that resembles a message or that might be from an intelligent civilisation. However, we need not actually restrict ourselves to stars within 45 light years.

Imagine that aliens from the Andromeda galaxy, for example, started sending out signals 2.5 million years ago. We would be receiving these now if they existed. The same applies to all of the other galaxies out there. Nor have we heard anything from those stars that have been targeted for close scrutiny by SETI investigators.

Table F. Distances of stars and planets

STAR	DISTANCE (light years)	PLANETS
Alpha Centauri B	4	1
Epsilon Eridani	10	1
Gliese 876	15	4
82 Eridani	20	3
Gliese 581	20	4
61 Virginis	28	3
55 Cancri	40	5
HD 69830	41	3
HD 40307	42	6

©Wikipedia/NASA.

Mention should be made of what has become known as the Wow! signal. This was a strong radio signal detected by a SETI researcher, Dr. Jerry Ehman, on 15 August 1977 (Images 99 and 100). The numbers and letters circled are a measure of the strength of the signal as compared to the background. In non-technical terms, the signal was 30 times stronger than any background noise, and lasted for the full time that the telescope was listening to that region of space, which was 72 seconds.

Image 99 is the actual computer printout on which Dr.Ehman wrote the word 'Wow!' Image 100 is a graphical display of the data which makes it easier to appreciate the sudden increase in the strength of the radio signal.

Unfortunately, in spite of repeated attempts since the signal was detected, it has never been heard again. So, much as many people would have liked this to have been a genuine part message from an alien civilisation, we have to assume that it was due to some other non-intelligent source. Even its

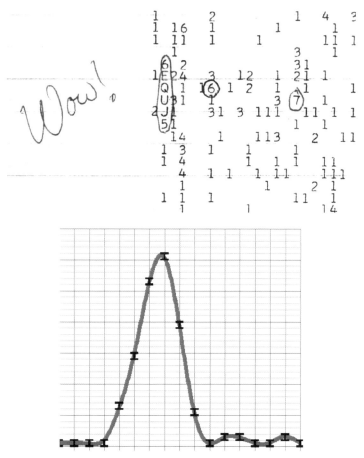

Images 99 & 100.Wow! signal printouts. ©*The Ohio State University Radio Observatory and the North American AstroPhysical Observatory.*

discoverer cautions against making "vast conclusions from half-vast data".

This is not conclusive evidence, of course. There may be planets around some of these stars with abundant life but none which has yet evolved sufficiently to build a radio telescope. All we can say is that at the present time there is no evidence of any extra-terrestrial life.

This lack of evidence, however, has not stopped some scientists from making bold claims. In 2010, a team of scientists at the University of California announced the discovery of an Earth-sized rocky planet orbiting a star known as Gliese 581.

Gliese 581 is about 20 light years from Earth, and the new planet, Gliese 581g, orbits its star in the Goldilocks zone where liquid water could exist. This made Gliese 581g the most Earth- like planet so far discovered.

The lead scientist, Professor Stephen Vogt, has stated, "personally, given the ubiquity and propensity of life to flourish wherever it can, my own personal feeling is that the chances of life on this planet are 100 percent."

That seems like a very rash statement. It is a potential candidate for life but that is about as far as one can go. Also, it should be noted that there is now doubt as to whether Gliese 581g actually exists and it no longer appears on NASA's list of confirmed exoplanets. In 2008, a high-powered radio signal containing 501 messages (look up 'A Message From Earth') was sent to planet Gliese 581c. It will get there in 2029.

A quote from Professor Stephen Hawking appears on the website of the National Space Centre in Leicester: "Given the thousands of Earth-like planets outside the solar system, on purely statistical grounds life almost certainly exists somewhere else."

It is my personal opinion that this is an unjustifiable conclusion since no information is available regarding the statistics of the emergence of life.

Remote Detection—Biomarkers

Radio signals are just one way in which alien life might announce itself. The most direct way of course would be an actual visitation, either historical with irrefutable evidence, or current. We have seen that there is no convincing evidence for either of these scenarios.

An alternative 'calling card' would be the presence of biomarkers on another planet or moon. A biomarker is a detectable chemical substance whose presence is indicative of a biological process, that is, of life. The usefulness of such a substance depends both on the probability of a living organism creating it and the improbability of a non-biological process creating it.

Space probes have visited every planet and many of the moons in our Solar System and methane (CH_4), a simple organic gas, has been detected in the atmospheres of most of the planets and many of the moons. It has been proposed as a possible biomarker but although it is produced by a variety of metabolic and hence biological processes, it can also arise from non-biological processes. Its presence therefore cannot on its own be used as an indicator of alien life. No definitive indicators of biological processes have so far been detected in our Solar System.

Exoplanets are far more difficult to study due to their extreme distances. However, spectroscopic analysis of starlight passing through their atmospheres during a transit of their star could allow the composition to be determined. Although this is currently at the limit of what is possible, the forthcoming James Webb Space Telescope, and others in development, should be up to the challenge.

For example, the observation of photosynthetic processes by its indirect detection of renewable oxygen in the atmosphere would be a powerful indicator of biological processes. Even more advanced future telescopes may be able to detect chlorophyll-like or other complex organic molecules on an exoplanet's surface.

More bizarre observations might also be possible. For example, a suitably advanced alien civilisation may have launched numerous solar panels into its atmosphere as a permanent energy source and such an array could perhaps be detectable from a space-based telescope.

Since travel to exoplanets is at best many decades away and in most cases far longer than that if at all, biomarkers may prove to be our best chance for detecting the presence of alien life, should it exist. And with several advanced telescopes due for completion within the next decade, an answer to this question could be possible within our lifetimes.

The journal *Astrobiology* (published by Mary Ann Liebert Inc., New York, USA) has many articles on this topic and is worth reading.

So, disappointing as it may be to some, there is currently no evidence of extra-terrestrial life and even the undeniably vast numbers of exoplanets do not justify us speculating on the likelihood of alien life. We may find it one day, or we may never find it because it is not there—there is just us. (See note 'Phosphine on Venus' at the end of the chapter.)

Extra-terrestrial life did exist but is now extinct

This is an interesting idea but unlikely ever to be provable. As we have seen in Question 3, life on Earth took about 1 billion years to get going, and then it took another 3 billion years or so for complex life to arise. What we call intelligent life, humans able to create technology, is a relatively recent development. So out of Earth's 4.5-billion-year history, the ability of its life forms to send messages to extra-terrestrial destinations can be measured in decades. Or we can say that life capable of extra-terrestrial communication has existed on Earth for approximately one 50 millionth of its existence.

What we do not know however, and may never know, is how long our civilisation, that is, all of humanity, will last. (It is the final term in the Drake Equation). That is not quite true, since we do have an upper limit based on the life cycle of the Sun (Question 9) of another 5 billion years or so. If alien life had arisen elsewhere in the universe, it may have started several billion years before ours did, and may also have succumbed to various extreme scenarios before life on Earth got going. If the alien planet's parent star had exploded, then any evidence of life would have been destroyed and would be undiscoverable forever. If the planet had suffered a major asteroid impact, or the life forms had become extinct or destroyed by some natural or artificial calamity, then remnants might remain to be discovered by future astronauts.

It is a bit like archaeologists coming across the remnants of an ancient civilisation, such as the ruins of Tiahuanaco, 3,870 m (12,700 ft) above sea level in Bolivia.

***Image 101**. Tiahuanaco ruins, Bolivia, c. 500 BCE. ©Author.*

Little is known about this lost civilisation since it left no written records. It probably arose around 500 BCE and continued to thrive for 1,000 years or so after which it died out. So, although we cannot deduce much about the inhabitants from these ruins, we do at least know that someone was there.

But what if...?

But what if one day we do actually receive an alien signal with irrefutable evidence that it is genuine. There would be all manner of implications, not the least being that we would then know for sure that life was not a one-time occurrence. We could then reasonably speculate that life is likely to be common throughout the universe. The implications would be immense. Scientific and religious pundits would have a field day as would conspiracy theorists. Just for fun, I have included an imaginary newspaper front page reporting such an event (Image 102).

And what about the next step? What if aliens actually do land on Earth? Will they come out of their spacecraft and demand "Take me to your leader!" If so, we shall be ready because in late 2010 the United Nations (UN) reportedly appointed a Malaysian astrophysicist, Dr. Mazlan Othman, as their Space Ambassador for Extraterrestrial Contact Affairs. It is not entirely clear whether this was a genuine report or not since it was denied by Dr. Othman who has in any case now retired from UN work. So, the job may still be available.

What questions would we wish to ask of our visitor? How would we communicate? Spoken language is unlikely to be helpful but perhaps a combination of gestures and drawings could be used. For example, we could point to the Sun and then draw it on paper surrounded by 7 planets and point to

Mysteries of the Universe 153

The Daily Star

www.dailystar.com THE UNIVERSE'S FAVOURITE NEWSPAPER Friday 10 October 2041 0.001 BITCOIN

ALIEN MESSAGE RECEIVED
by Peter Altman, Staff Reporter

Andromeda Galaxy

What the message might look like

1100101010100101010100
0101010101000010010011

The Andromeda Galaxy is 2.5 million miles from Earth and can just be seen with the naked eye on a clear night as a faint smudge. Three days ago, NASA announced that its orbiting radio telescope had detected a series of radio signals from the direction of Andromeda in the form of a repeating series of binary digits. No further information was released except to confirm that the signals could not have originated from a natural source.

There is intense speculation as to the content of the message. It has to be remembered however that the message would have been sent 2.5 million years ago so that the senders would be long dead – probably. Journalists and scientists are inundating NASA for more information and for sight of the message but it is being kept secret until some idea of the likely content has been deciphered. This may however turn out to be impossible.

Image 102. Imaginary newspaper report announcing the receipt of an alien radio message from the Andromeda galaxy. ©Author.

the third one and then to the ground to indicate our planet. A similar device was used on the Pioneer plaques to indicate to any intercepting aliens where the Pioneer spacecraft had come from.

Then, giving the paper and pencil to the alien would hopefully encourage him/her/it to create a similar drawing of his/her/its origins. Other drawings and gestures may then permit more complex questions to be answered.

Again, just for fun, here are some questions that we may wish to ask (Table G).

Table G. Alien registration form

#	Question
1	Where have you come from?
2	Why did you come here?
3	How did you get here?
4	How long did it take?
5	How far can you travel?
6	What other life forms exist on your planet?
7	Have you found Life elsewhere?
8	Do you suffer from illnesses?
9	Do you have wars?
10	How long do you live?
11	How do you communicate?
12	Do you worship Gods?
13	What do you eat and drink?
14	How do you reproduce?
15	Can you travel through time?
16	What powers your spacecraft?
17	How old is your civilisation?
18	May I take some Photographs?
19	May I have some articles as souvenirs?
20	What is the basis of your biochemistry?

Note: Phosphine on Venus

Phosphine gas is produced by some microbes on Earth and has now (September 2020) been detected on Venus. Three things need confirmation. 1. *There really is phosphine in Venus's atmosphere. 2. It is produced by living organisms. 3. The organisms are not a contamination from any of the 38 probes that have been sent to Venus since 1962.* That would be a stunning finding. If Life has also developed on our nearest planet, then it is reasonable to assume that it is likely to be common throughout the universe.

THIS IS AN UNANSWERABLE QUESTION

Best Guess Answer

Life, as we must consider it, needs to arise on a rocky planet (or moon) located within the habitable and abiogenesis zones of its star, probably with a large moon and a magnetic field, and contain the necessary ingredients for life. Planets seem to be common around other stars so it is likely that many stars would have the right sort of planet in the right position for life to develop.

Moons and exomoons

Although way beyond the habitable zone of the Sun, it has been suggested that some of the moons of Jupiter and Saturn could harbour primitive life. This is because they are volcanically active, probably due to vast tidal forces from their giant home planets, and the heat generated by these forces could allow for large oceans of liquid water beneath their surface. It is conceivable therefore that planets and moons far beyond a star's habitable zone could still have sub-surface temperatures high enough for liquid water.

The JUpiter ICy moons Explorer (JUICE) probe is due to launch in 2022 and will reach the Jovian system in 2030 to start searching for signs of life.

There is also great interest in the possible future discoveries of exomoons—moons of exoplanets, although 6 of which have been announced to date (2020).

If our Solar System is anything to go by (eight planets, several dwarf planets and over 200 known moons), exomoons may be very common indeed.

There are likely to be thousands of stars within a distance from Earth that gives enough time for an alien civilisation to intercept radio broadcasts from Earth and send coded signals back to us. There is also the possibility of one-way signals sent millions or even billions of years ago from anywhere in the universe which could have reached us by now. There is no evidence of any such signals having been received.

We have no information at all as to how likely it is that life will develop on a suitable planet and calculations attempting to quantify these probabilities are invalid due to the lack of suitable data. Life may be so unlikely that it has only happened once or it may be extremely likely so that it has developed many times. We just do not know and we also cannot speculate on its likelihood. All we can say is that there is no current evidence for any intelligent, or indeed any other, life anywhere else in the universe. That of course does not mean it is not there—it just means we have not found it.

There is also no evidence that life has arisen on Earth more than once. The question of whether life had ever existed elsewhere in the past but is now extinct is probably forever unanswerable.

We can consider arguments for and against the presence of alien life.

Arguments for

1 Vast numbers of planets and moons in our, and presumably all other, galaxies (although this also

works both ways—see point 4)
2 Presence of organic materials in meteorites
3 Plenty of time for life to develop (this also works both ways—see point 4 below)
4 Possible detection of phosphine on Venus

Arguments against

1 No radio signals indicating an intelligent source have been detected in over 50 years of listening
2 No biochemical changes indicating living organisms have been detected on other planets or moons
3 No credible evidence for alien artefacts or visitations
4 Time. The Sun is about 5 billion years old whereas some stars in the Milky Way and in many other galaxies are more than 10 billion years old; the oldest are even 13 billion years old. Alien life would therefore have had an extra 5 to 8 billion years to develop on some of the vast numbers of planets of these much older stars compared to the time available in our own Solar System, yet there is no evidence for its existence. The extra billions of years of evolution would have given time for aliens to develop technologies to explore their home galaxy and even others. This is the Fermi Paradox enumerated by physicist Enrico Fermi in 1950—given the vast numbers of planets and time for billions of years of evolution, the universe should be teeming with life.

If that is the case, then where is everybody? We should have heard from them by now. The simplest answer is that they do not exist. This is known as the Rare Earth hypothesis. Basically, it states that the emergence of intelligent life on Earth required a very unlikely combination of specific astronomical, geological, physical and biochemical conditions which, taken together, are so improbable that they have happened only once and are unique to the Earth.

Many people, both lay and professional, claim that life is likely to be common throughout the universe because of the vast numbers of planets that almost certainly exist. I subscribe to the Rare Earth hypothesis. Since we do not know how likely it was that life arose on Earth, we cannot assume that even 6,000,000,000,000,000,000 planets are sufficient to make it more likely than not. If a roulette wheel with an unknown number of numbers stops at number 10 say, then even if there are 6,000,000,000,000,000,000 roulette wheels spinning, we cannot say that it is more likely than not that number 10 will come up again on one of the wheels because with an unknown number of numbers, it is impossible to calculate the odds.

This chapter contains a calculation about tossing a coin, and shows that the chances of heads coming up 62 times in a row is about 1 in 6 million million million, equivalent to one very unlikely random event occurring on just one planet in the entire universe. This demonstrates that a very mundane event, in this case, 62 consecutive heads can be unimaginably unlikely.

Make it four more consecutive heads and the outcome is 16 (2^4) times less likely. Contrast these consecutive heads with the emergence of a living organism. That sounds far less likely. It certainly happened once because here we are, but more than once?

So, where does this leave us? Clearly, we cannot give a definitive answer as to whether alien life exists. That question will not be answerable until evidence for it is found. This is a science book, and the evidence for UFOs, alien visitations in the distant past, and modern alien abductions, is just not robust enough to be taken seriously.

Although we can never say for sure that alien life does not exist (since in this case proving a negative is impossible), we need to decide upon a best guess answer based on what is known now.

There are basically two schools of thought. One, probably held by most people, is that the universe is likely to contain huge numbers of Earth-like planets so that the emergence of life on at least some of them is very likely.

The other posits an opposing view—if the universe is indeed teeming with life then why have we not detected any evidence of it? (The Fermi Paradox). Why? Because it is not there. Why is it not there? Because its formation on Earth was the result of an extremely unlikely combination—a Perfect Storm of geological, biochemical, astronomical, climatological and probably many other factors that came together in just the right place at just the right time. And as discussed earlier in this Question, it is also very probable that the very rare presence of a large moon in comparison to the size of the Earth was an important chance event, resulting in a stabilising effect on Earth's climate. Finally, we should also consider the probable origin of all the water on Earth as due to a chance encounter with an asteroid or comet.

This juxtaposition of conditions is so unlikely that even 6×10^{18} potential Earth-like planets are not enough for it to have happened more than once. This is the Rare Earth Hypothesis and it is the view that I personally hold. Based on current knowledge therefore my best guess answer is that alien life does not exist.

Q6

COULD WE TRAVEL TO OTHER STARS AND GALAXIES?

Will we ever travel to the stars or to other galaxies? Could it be done?

LEAVING THE SOLAR SYSTEM

Let us first consider un-manned journeys. Out of all the space probes that have been launched, several have or will leave the Solar System and continue their journeys into interstellar space until they are disturbed by accident, impact, or retrieval by an alien civilisation (Table H). The latter scenario would provide, for that civilisation, irrefutable evidence of alien life.

The last column shows the general direction in which the probes are travelling. Apart from New Horizons, which encountered a Kuiper Belt object (Ultima Thule, now renamed 486958 Arrokoth) in 2019, none have been deliberately aimed at any particular astronomical object after their primary mission was completed and, as stated above, all will continue to travel until and unless their journey is interrupted.

Table H. Details of space probes

PROBE	LAUNCH	LAST CONTACT	CURRENT DISTANCE (2020)	GOING TOWARDS
Pioneer 10	1972	2003	10 billion miles	Taurus
Pioneer 11	1973	1996	9 billion miles	Scutum
Voyager 2	1977	active	11 billion miles	Pavo
Voyager 1	1977	active	14 billion miles	Ophiucus
New Horizons	2006	active	4 billion miles	Sagittarius

Looking at these figures you may think that the Pioneers and Voyagers, which have travelled around 10 billion miles, have gone a long way. They certainly have, but only in relation to Earth terms. In astronomical terms they have barely gone anywhere; 10 billion miles is less than 0.2 percent of the way to Proxima Centauri, the nearest star to the Sun. At their current speed, these spacecraft would take around 100,000 years to reach Proxima Centauri if they were headed in that direction.

Given the vast distances between stars, an accidental encounter with a star or planet seems exceptionally unlikely. Even if that should happen, such encounters would be in the range of tens of thousands to billions of years away so no information of such an event could ever be transmitted back to Earth.

What about manned missions? It took the New Horizons spacecraft, launched in 2006, nine years to reach Pluto, travelling at an average speed of about 60,000 kph (37,000 mph). If the technology could be upgraded to take astronauts, then we would be talking about a 20-year round trip, which is possibly feasible if anyone was prepared to be away that long. However, although journeys to the Moon became almost routine in the 1970s, upgrading the technology from a five-day to a several months' journey (to Mars) really is a giant leap that is probably still a decade or more away. Longer manned missions to the outer Solar System, while possibly doable in a few decades, will also face the more mundane difficulty of funding.

Disregarding the funding issues, crewed interstellar missions seem impossible without a really significant breakthrough in rocket propulsion technology. If, as above, we assume that a 20-year round trip for an astronaut crew is the maximum practical time for a space voyage, and that you would not go all that way without staying for a while, we have to say that about eight or nine years for a one-way journey to Proxima Centauri is the longest time available. That means travelling at half the speed of light, over 480 million kph (300 million mph). Alternative options would be for a slower journey time where the crew would live and die on board and a future generation would take over the mission, or where the crew would be in suspended animation with an autonomous spacecraft.

These scenarios are impossible with foreseeable technology. There are various ideas on the drawing board however.

Spacecraft powered by nuclear pulse propulsion engines have been proposed and researched. These engines would develop millions of tons of thrust by having directed nuclear explosions driving the craft forwards. The aim was to design a spacecraft that could reach Proxima Centauri in less than a crew member's or mission control member's working lifetime (taken as 50 years).

Constant acceleration drives could also make interstellar journeys feasible. In contrast to current modes of rocket propulsion where there is an initial engine burst to get the spacecraft under way followed by the remainder of the journey in 'coasting' mode, a constant acceleration drive is on all the time, continuously accelerating the spacecraft. In other words, the spacecraft just goes faster and faster, eventually approaching the speed of light.

With a constant acceleration of 0.5g (g is the acceleration due to gravity on Earth, about 9.8 m (32 ft) per second, that is, every second an object falls it goes 9.8 m (32 ft) per second faster than during the previous second), a spacecraft would approach the speed of light in about one year. During this time, it would have travelled about half a light year in distance.

As an approximation, we can say that the journey time with such a constant acceleration drive would be the distance in light years to the destination plus one year. A journey to Proxima Centauri, just over four light years away, would therefore take about five years.

A novel type of propulsion system, known as the Electro-Magnetic Drive or EmDrive, has received some press coverage.

The EmDrive works by converting electric power into forward thrust. The drive accelerates a craft gradually but continuously, thereby making it go faster and faster—a constant acceleration drive as mentioned earlier. Its inventor, British aerospace engineer Roger Shawyer, has calculated that a spaceship would take ten years to reach two-thirds of the speed of light—about 450 million mph.

The effects of relativity will be relevant meaning that the elapsed time frame for the crew will be shorter than that for observers back on Earth. For example, a journey time from the crew's point of view of five years might appear to observers on Earth as having taken fifty years. So, if the crew were away for ten years, the elapsed time back on Earth might be one hundred years.

The time difference would be much greater for longer journeys and could run into thousands of years. The crew will, in effect, have travelled into the future.

It has been suggested that a mission which cannot reach its destination within fifty years should not be started at all. Instead, the money should be invested in designing a better propulsion system. This is because a slow spacecraft would probably be beaten by another mission sent later with more advanced propulsion, making the first mission a waste of resources.

Intergalactic travel with journey times measured in millions of years remains in the realms of science fiction. Such journeys will require more than just faster rocket motors or sleeping crew members in an autonomous spacecraft. Distances measured in thousands and millions of light years could never be accomplished by straight line travel even at speeds close to that of light since crews would never survive the journey. Does that mean they will never be done?

There could be some theoretical ways around this.

TRAVEL AT FASTER-THAN-LIGHT SPEEDS
Is this possible?

As outlined in the section 'Size Matters' in Question 1, the space between galaxies is expanding and this expansion can be at super-luminal speeds. Because space is not a material object, it is not subject to the light-speed speed limit.

But what about material objects travelling faster than light? There are many entries on Google under the term 'faster than light travel'. Several of these are from scientists who have used advanced mathematics and physics to demonstrate, on paper at least, that this may be possible without infringing the known laws of physics.

Research workers at the NASA Johnson Space Center believe that it may be possible to harness the known expansion and contraction of space to power a spacecraft to superluminal speeds. Since it is known that space can expand at faster-than-light speeds, Dr. Harold White, one of the researchers, has said, "If Nature can travel at warp speeds, perhaps humans can as well."

An intriguing experimental result was announced towards the end of 2011 from the CERN laboratory in Geneva. Briefly, the scientists had apparently accelerated a beam of neutrinos—a type of fundamental particle—to a speed greater than that of light. The neutrinos were sent on a 732-km (455-mile) journey from one laboratory to another. The journey, at light speed, should have taken 0.0024 seconds but the neutrinos arrived early. However, not very early—only by 0.0000006 seconds. That is not a lot in layman's terms but as far as physicists are concerned it is highly significant.

In 2012, it was announced that this result was an error due to equipment errors in the original experiment. Repeated experiments had shown that neutrinos did indeed travel at the speed of light and not any faster. Shame!

If we could travel at one million times the speed of light, then we could go from one end of our galaxy to the other in about a month, and could reach the Andromeda galaxy in 2.5 years. It is a very long journey indeed, however, from some equations in a research article to a working spacecraft.

Even if this could one day be achieved, there are huge potential problems with such journeys. The fact that the Andromeda galaxy, for example, is 2.5 million light years away means that we are seeing it now as it was when the light left, that is, 2.5 million years ago. We do not even know if Andromeda is there now. The astronauts could arrive to find that it has gone. Moved, exploded—who knows? "Sorry guys, we're too late. Let's go back home."

TRAVEL THROUGH WORM HOLES IN SPACE

A worm hole is a purely theoretical concept. Imagine space as the surface of a balloon. To travel from one point on the surface of the balloon to another, you have to make a curved

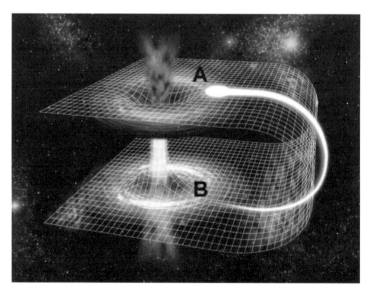

***Image 103**. Worm hole. Space is represented as a curved surface. A spacecraft going from A to B along the curve would clearly have a longer journey in both time and distance than a spacecraft going through the worm hole. ©www.dreamstime.com.*

path. Some physicists have suggested that space is curved like this. If that is so, then a worm hole is like a path taken through the balloon, rather than over its surface.

This path is of course shorter than the surface one so the journey time would be faster. As the curvature becomes more extreme, so the travel distance becomes shorter (see Image 103 on the previous page). But it is only a theory.

It is a brave person who says that something is impossible. Look at all the technology available today and imagine what people living 100 years ago would think if they could see it now. It is a mistake however to think that everything develops and will eventually happen. Here is an example.

The first controlled flight of a powered aeroplane was made by Orville Wright at Kitty Hawk, North Carolina, USA, on 17 December 1903.

Seventy three years of development resulted in the first flight of Concorde. This aircraft had a range of 7,250 km (4,500 miles), could reach twice the speed of sound and

***Image 104**. Orville Wright's first flight with Wilbur alongside, 1903. ©NASA.*

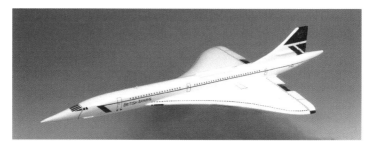

Image 105. Concorde. ©Wikipedia.

cruise at 18,250 m (60,000 ft), high enough to see the Earth's curvature.

It does not always happen like that though. In 1969, Apollo 11 landed on the Moon.

What would have been the expectation, in 1969, of how manned space exploration would develop over the next 50 years? Manned colonies on the Moon? Regular trips to Mars? And what has happened?

Not much.

Image 106. Buzz Aldrin stepping onto the Moon from Apollo 11 in 1969. ©NASA.

Why? The main reason is cost. Even Concorde was so expensive to develop (about £20 billion in today's money) that it became a shared project between the UK and France.

The Apollo space program cost, in today's money, close to $200 billion. Public support is essential when governments wish to spend money at that level, and after seven Apollo missions to the Moon, that support disappeared.

More complex projects, such as manned planetary missions and lunar colonisation missions, would be vastly more expensive, and with more immediate problems on Earth to deal with, the resources to develop these space missions are just not available.

What has happened however is that various private companies, including Sir Richard Branson's Virgin Galactica, Elon Musk's Space X and Jeff Bezos's Blue Origin, have begun to develop their own rockets for planned trips to the Moon and Mars. On 18 September 2018, Space X announced that probably in 2023 they will be taking a Japanese billionaire, Mr. Yusaka Maezawa, on a trip around the Moon together with six artists to capture the experience. The cost was not divulged but is likely to be in the region of many tens of millions of dollars per seat. It seems probable therefore, that space exploration, at least within our own Solar System, may actually get going during the next few decades.

But what about interstellar and intergalactic journeys by faster-than-light travel or worm holes or by other yet-to-be discovered technologies? Obviously, we cannot say that these things will never happen but it does seem likely that such developments are at least hundreds if not thousands of years away, if indeed they ever happen at all.

It is a shame really. We may be forever trapped by the laws of physics to remain within the confines of our own Solar

System. There could be and probably are unknown wonders within the universe that we will never visit.

> **THIS IS AN UNANSWERABLE QUESTION**
>
> **Best Guess Answer**
>
> Interplanetary manned travel within our Solar System is doable with appropriate financial support. Although most of the proposed missions have so far been government-funded, there is increasing interest amongst a small number of private companies. The first private citizen to visit the Moon has already booked his seat.
>
> Interstellar travel, at least to our closest stellar neighbours, may become theoretically possible with foreseeable technology but would either require many generations of astronauts to complete the journey, or a system of deep suspended animation for the entire crew, or the development of nuclear-powered constant acceleration drives.
>
> Longer distance interstellar and all intergalactic travel where the distances are measured in thousands and millions of light years would require either faster-than-light travel or autonomous spacecrafts capable of operating continuously for many millions of years. Unmanned missions such as those which have explored the Solar System would certainly be attempted first.
>
> They would require completely autonomous spacecrafts since the travel time for signals to and from Mission Control would be far too long.

Such a mission with a living crew in suspended animation seems unimaginable. However, if self-replicating robots could be developed, it might be possible to colonise a galaxy in several million years. Given that the Sun still has a few billion years of life left, future technology might make this possible.

This then relates to the question of whether alien life exists (Question 5). Alien civilisations, if they exist, could have had millions or even billions of extra years of development compared to our own, especially if their star is older than the Sun. If that has not given enough time to colonise a galaxy with robots, which as far as we know it has not, then presumably the aliens just do not exist.

My best guess answer is that human interstellar travel to our nearest neighbours may be possible but destinations further afield within our galaxy and all intergalactic travel are not feasible. But, in a distant future, robotic visitations may be achievable.

Q7

IS TIME TRAVEL POSSIBLE?

Time, the period between two events, is usually called the fourth dimension and travelling backwards and forwards through time is a concept beloved of science fiction writers. There are many examples, including H G Wells's novel, *The*

Image 107*. Can we travel through time? Poster for* The Time Machine *(1960), an MGM film. ©Wikimedia Commons.*

Time Machine, published in 1895 and filmed in 1960, and the film *Back to the Future* which was released in 1985. But is time travel possible in real life?

TRAVELLING BACK IN TIME

It is certainly possible to look back in time. Since the speed of light is finite, it takes a certain amount of time to travel from its source to your eyes. When you glance at your watch to see the time, your eyes are about 30 cm (12 in) from your wrist. Light takes about one billionth of a second to travel 30 cm (12 in) so you are actually seeing your watch, and the time, as it was one billionth of a second ago.

Whenever we look at something we are always looking back in time. It is, in fact, impossible to look at something and see it as it is now.

This does not make much difference in practice though. When you look at the Sun, you are seeing it as it was 8 minutes and 20 seconds ago because that is how long it takes for the Sun's rays to travel to the Earth. If the Sun had suddenly disappeared just after you looked at it, you would have no knowledge of this for 8 minutes and 20 seconds. It would look and feel exactly the same due to the light and heat that had already left it before it disappeared.

The light from the Andromeda galaxy takes 2.5 million years to reach the Earth, so when we see it as a smudge in the sky, or as a spiral galaxy through a large telescope, we are actually looking 2.5 million years into the past.

If we could one day develop telescopes powerful enough to see detail on stars and planets within the Andromeda galaxy, then we would be observing in real time events that took place 2.5 million years ago.

If an advanced alien on a planet in the M61 spiral galaxy,

which is about 65 million light years away, looked at the Earth now with a sufficiently powerful telescope, he/she/it would be able to see dinosaurs roaming around as depicted in Image 108. That is because the light rays making up this image left their source 65 million years ago and have taken that time to reach the observer in the M61 galaxy.

But what about travelling back in time rather than just looking? The so-called Grandfather Paradox is one of several thought experiments that make it seem that travelling backwards in time is impossible. Imagine that you were able to travel back in time and that you met and killed your own grandfather before he met your grandmother. This would mean that your father, and therefore yourself, would never have been born. The only way to resolve this paradox is to say that travelling back in time is impossible.

Well, not quite the only way. You could also argue that

Image 108. *View of Earth now through an alien telescope on a planet 65 million light years away. ©Author.*

travelling backwards in time is actually possible but that it is impossible to interfere with history. What has happened has happened and cannot be changed. That does not seem to be a very satisfactory argument however.

Some additional evidence, if that is the right word, against the possibility of backwards time travel is the following argument—if it was possible to travel back in time, then why are there no time travellers here from the future?

It has been suggested that such travellers are and have been here but disguised as present-day people so that we do not recognise them as from the future. That makes no sense—if you had travelled back in time you would surely wish to make contact and distribute your knowledge of what was to come.

Based on current opinion therefore it seems that travelling back in time is not possible.

TRAVELLING FORWARD IN TIME

The perception of time can be speeded up by hibernation or suspended animation where body temperatures and metabolic rates are reduced. Time will seem to have passed more quickly for the organism that wakes from hibernation or suspended animation, or even from an anaesthetic or from sleep.

However, this is not really time travel since the same time would have passed whether suspended animation, etc. existed or not. It is only the perception of time that has altered.

TIME DILATION

One of the consequences of the Theory of Relativity is that time slows down as speed increases. This is known as Time

Dilation. Although the effect is present at all speeds, it only becomes significant at speeds close to that of light.

Time Dilation refers to the slowing of time as speed increases. You are not going to notice this however as you drive along the motorway since the effect only becomes apparent at speeds close to the speed of light. Its effect can be estimated from the following graph, which shows the percentage that time is dilated (slowed down) at different speeds relative to that of light.

For example, travelling at 90 percent of the speed of light would result in a time dilation (slowing) of about 55 percent. In other words, a ten-year journey at that speed would seem to the travellers like a four and a half-year journey. The effect becomes even more pronounced at speeds closer to that of light, so that at 99 percent of the speed of light, the ten-year journey would seem like about one and a half years. At a speed of 99.9999 percent of the speed of light, the ten-year journey would take the travellers just five days!

This effectively means that the travellers have travelled into

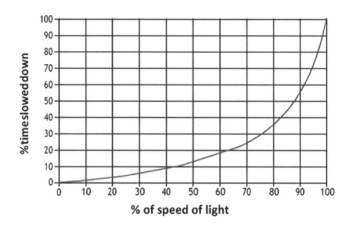

Image 109. *Time Dilation—showing the effect of speed on elapsed time.*

the future since upon their return to Earth at the same speed, the total elapsed time for the 20-year journey (for people on Earth) would have taken (for the travellers) only ten days.

Astronauts inside a spacecraft travelling at close to light speed, if that ever becomes possible, would find that everything in and on the craft showed less elapsed time than that measured by mission control and others back on Earth. In other words, time slows down for them.

They could reach far-off stars and galaxies well within a lifetime and find that millions of years had passed when they returned to Earth. This is the basis of the *Planet of the Apes* story.

Alternatively, they could perhaps just whizz around the Earth rather than travelling anywhere. As long as they were going fast enough for long enough, they could land after say ten of their years to find that maybe 1,000 years had passed on Earth. It is a one-way journey though, since there would be no way of travelling back to their starting date.

Time dilation has been tested experimentally by flying atomic clocks in aircraft and showing that there is a difference in elapsed time between these travelling clocks and similar atomic clocks back on Earth. Also, atomic clocks on GPS satellites circling the Earth need corrections due to time dilation. The differences are small though, of the order of millionths of a second a day.

The Russian cosmonaut Sergei Avdeyev has spent a total of over two years on the Mir space station travelling at about 27,000 kph (17,000 mph). Due to time dilation, it has been calculated that Avdeyev is 0.02 seconds younger than he would have been had he not travelled in space.

Worm holes, those theoretical features of spacetime that were mentioned in Question 6, could perhaps provide a pathway between the present and the future. It has to be said however, that this is currently much more in the realm of science fiction than science fact.

Whereas it seems that time travel into the past is not possible, there is at least the physical principle of time dilation that could, if the required near-light speeds are ever achieved, perhaps one day enable time travel into the future. It is likely to be a one-way journey though.

If constant acceleration drives could ever be developed (see Question 6), then speeds approaching that of light could perhaps be attained. This means that a journey to Proxima Centauri, just over four light years away, would take about five years to complete.

As explained in Question 6, because of relativity, the elapsed time frame for the crew will be shorter than that for observers back on Earth. The journey time from the crew's point of view of five years might appear to observers on Earth as having taken 50 years. The crew will in effect have travelled into the future because their elapsed time of five years there and five years back, and say a one-year stay, will mean that they have been away for eleven years. Back on Earth however, over 100 years may have passed.

We can end this section with a rather bizarre notion. We have seen that time travel into the past seems to be impossible but, if sufficiently fast speeds can be attained, time travel into the future is theoretically possible. So, if one day we could travel into the far distant future, then we might arrive at a time when the problems of travel into the past had been solved, and we could then go back to our time of origin. However, if the problem of backwards time travel was still insoluble, we would have to keep going forwards until we found ourselves in a time when we could go back. Bit of a risk really.

THIS IS AN UNANSWERABLE QUESTION

Best Guess Answer

Time travel can be either into the past or into the future. Conventional time travel as depicted in science fiction usually involves a transport device that can carry its occupants backwards and forwards in time. Travel into the past seems to be impossible since it is difficult to get around thought experiments such as the Grandfather Paradox. Additionally, if it were possible, we might have expected numerous visitors from the future but there is no evidence of such visits.

Travel into the future is allowed by the Theory of Relativity due to the effect known as time dilation. As an object's speed increases, the passage of time on and in the object slows down relative to that for observers on Earth. A crew travelling at near-light speed for say ten years would find that many hundreds of years had passed upon their return to Earth.

This is not quite the same as the putative machines of science fiction since it would likely be a one-way journey with no prospect of returning to the time of origin, unless of course they arrived in a distant future where the problems of backwards time travel had been overcome.

Time travel as per science fiction is probably impossible. One-way journeys into the future are theoretically possible, although construction of such transport devices is likely to be thousands of years away, if at all.

Q8

COULD WE LIVE LONGER?

It is a fair assumption that everyone would like to live longer but in good health. Still fit and well at 90 is preferable to bed-ridden and semi-conscious at 100. Medieval alchemists spent hundreds of years searching for the mythical Elixir of Life, a medical potion that was supposed to confer eternal youth. They never found it. Can modern science and medicine do any better?

Before we get to ways of possibly extending our lifespan, we need to consider how long we can expect to live without any intervention. This is known as the life expectancy and it varies a great deal according to country, with the poorest countries not surprisingly having the lowest life expectancies. This is because poverty goes hand in hand with poor nutrition, poor sanitation and poor health care.

Table I shows the life expectancy at birth for the top and bottom ten countries, plus the UK and the USA (UN data from 2016). The full table can be found on Wikipedia under 'Life Expectancies by country'.

Life expectancies, especially in the more developed countries, have steadily increased in recent years due to better awareness of important lifestyle factors as well as the introduction of new medications.

Table I. Life expectancies by country

SL	COUNTRY	LE (years)
1	Japan	83.7
2	Switzerland	83.4
3	Singapore	83.1
4	Australia	82.8
5	Spain	82.7
6	Iceland	82.7
7	Italy	82.7
8	Israel	82.5
9	Sweden	82.4
10	France	82.4
20	United Kingdom	81.2
31	United States	79.3
174	South Sudan	57.3
175	Cameroon	57.2
176	Somalia	55.0
177	Nigeria	54.5
178	Lesotho	53.7
179	Ivory Coast	53.3
180	Chad	53.1
181	Central African Republic	52.5
182	Angola	52.4
183	Sierra Leone	50.1

Source: Data modified from United Nations 2016.

Recent government figures show that 17 percent of the UK population can expect to live to 100. But what if we want to do better than this? What if we want to live for 200 years or even longer? Is this possible?

There are accounts of bacterial and other spores being

extracted and revived from mineral deposits and amber after hundreds of millions of years. Such organisms are able to survive in a state of deep suspended animation for extraordinary amounts of time. The oldest living things ever recorded are some bacteria, *Bacillus permians*, which were revived after being found embedded in salt crystals in New Mexico in 2000. They were dated as being 250 million years old.

Some pine and spruce trees have been found to be nearly 5,000 years old. Jellyfish are able to regenerate damaged or diseased parts of their bodies and can be considered immortal, and, unless eaten by prey or severely damaged by accident, can theoretically live forever. Animals struggle to reach 200 years and the oldest human to have lived was Jeanne Calment, a French lady who died in 1997, aged 122 years and 164 days. It should be noted however that doubt has been cast on this claim, suggesting that the lady who died in 1997 was actually Madame Calment's sister Yvonne who assumed her mother's identity when she died in 1934 to avoid paying inheritance tax and death duties. See https://www.smithsonianmag.com/smart-news/study-questions-age-worlds-oldest-woman-180971153/#jmLTsLVVycP55Mw4.99 for more details.

Whatever the true age of Jean Calment, there are certainly people who have lived into the 110s. Could this be improved upon? Could we live, if not forever, for hundreds or even thousands of years? We can consider various options, some available now and others maybe in the future.

SUPERCENTENARIANS

People aged 110 and above, known as supercentenarians, seem to have differences in their immune systems compared to those who die at more normal ages. Recent research has shown that supercentenarians have a higher resistance to the usual lethal diseases of old age and tend to live their lives in general good health.

HEALTHY LIFESTYLE

More people could certainly live longer if they adopted a healthier lifestyle although this is unlikely to result in significant overall increases in lifespan. In any case, even though it is widely known that exercise and a diet low in fats and sugar are both conducive to maintaining good health, in many countries significant percentages of the populations are classed as obese. The problem is that fatty and sweet foods taste good and most people are not keen on exercise. Watching television with a hamburger and fries is more agreeable than going on a run and then having a salad.

THE AGEING PROCESS

Is it possible to manipulate the ageing process? There is a very rare genetic disease known as progeria which affects about one in eight million people. Those born with this condition age much more quickly than normal people and usually die in their teens due to diseases of old age, such as hardening of the arteries and heart failure. As doctors and scientists learn more about how this disease speeds up the ageing process, they may also find ways of slowing it down.

TELOMERES

Research work on telomeres may prove rewarding. Telomeres are the end bits of chromosomes rather like the metal tips of shoelaces. Both serve similar functions in that they prevent the ends from fraying. Telomeres protect the ends of

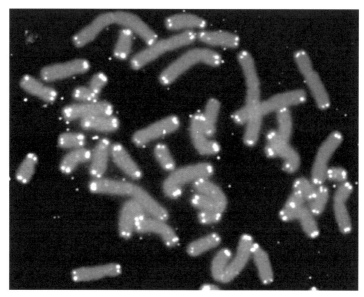

Image 110. Human chromosomes showing telomeres as white spots. ©Wikimedia Commons.

chromosomes during the duplication process and become shortened after each cell division.

Eventually, they become so short that they can no longer function properly, and the cell cannot duplicate so it dies. This is possibly either a cause or a consequence of ageing.

There is a limit as to the number of times a cell can divide before it stops, presumably because the telomeres have become too short. For human cells, this is between about 60 and 80 times. Telomeres can be lengthened by an enzyme, telomerase, and it has been suggested that increasing the length of the telomeres might enable cells to divide for longer and the ageing process would be slowed down. Interestingly, telomeres from patients with progeria are shorter than those from normal people of the same age.

In 2011, a Spanish company, Life Length, announced the

availability of telomere length measurements and a calculated estimate of life expectancy. This will not, of course, tell you if you will be run over by a bus.

If it turns out that a shortened telomere length really is a cause of ageing (this is not proven yet—shorter telomeres could be a result of ageing rather than a cause), then telomerase may provide a mechanism whereby the ageing process really could be slowed down. This could be a modern-day Elixir of Life.

It is likely to be a long way off though since even if a way of lengthening telomeres in humans can be found, there would need to be rigorous studies to show that it is safe to do so. One potential risk would be cancer. Cancer cells divide uncontrollably and have longer telomeres than normal cells.

CALORIE RESTRICTION

It has been known for many years that laboratory animals live longer on a calorie-restricted diet. In 2010, it was reported that the lifespan of mice could be extended by about 12 percent when they were fed with water containing some essential amino acids. They were also stronger and had more stamina than the control group. This is yet to be tried on humans but if proven, could be a cheap and acceptable way of increasing lifespan and better health in old age.

Recently there have been some small clinical trials on intermittent fasting where calorie-restricted diets are followed for two non-consecutive days in every week. More data is needed to see if this works although early results have shown some benefit for both weight loss and an improvement in cholesterol levels. However, even if it were true that the human lifespan could be extended in this way by say ten years, it is doubtful whether anyone would take this up. Most people enjoy their food too much to give it up for 50 years on a promise of a few more years of life.

RAPAMYCIN

Rapamycin is produced by a strain of bacteria and was discovered in a soil sample from Easter Island in the 1970s. It is used clinically to suppress the immune system and prevent rejection in patients undergoing organ transplantation.

In 2006, it was discovered that the drug could extend the lifespan of yeast cells. Further work in the following years then showed that the lifespan of mice could be extended significantly by feeding them rapamycin. Studies on dogs and monkeys are underway.

It is not known whether rapamycin will have lifespan-enhancing effects in humans or whether it is even safe to take. However, much has been learnt about the drug's mechanism of action and which biochemical systems it targets. This may then enable the synthesis of more effective and safer drugs in the future.

SENOLYTICS

Senolytics are drugs that destroy old and worn out cells which can accumulate in tissues and cause damage. They are known to cause a variety of age-related diseases such as Type 2 diabetes, cancer, Alzheimer's disease, Parkinson's disease and many others. Various experiments have shown that if senescent cells are removed from old laboratory animals, they become rejuvenated.

Not surprisingly therefore, this has become an active area of research. It should be emphasised however that the aim here is not so much to increase lifespan but to increase health in old age. No one wants to create large numbers of centenarians all suffering from many of the usual age-related diseases; the target is to have, say, 85-year-olds in good health. That is what, hopefully, senolytics may be able to deliver.

CRYOPRESERVATION AND RE-VITALISATION

Cryopreservation, where a body is frozen immediately after death in the hope of revival when future medical advances can treat the cause of death, has become popular, especially in the US. The body, or sometimes just the head, is frozen and then stored in liquid nitrogen at -196°C and returned to normal temperatures when the revival process begins.

One of the major companies providing cryopreservation is the Cryonics Institute based in Michigan in the US. As of August 2019, it has 177 frozen patients and a number of frozen pets. Nearly 1,800 have taken out membership at an annual cost of around $1,000; the actual cryopreservation process costs about $30,000. Most scientists and doctors are sceptical as to whether revival will be possible.

In September 2018, there was a news report that the son of a man who was frozen after death by an American cryopreservation company had found out that only his father's head had been preserved, the rest of his body having been cremated. He is now suing the company for $1 million.

In May 2016 a US Biotech company, Bioquark Inc. in collaboration with an Indian Biotech company Revita Lifesciences, was given permission to begin recruiting a small number of clinically dead patients and attempt to revive their central nervous systems. This was the Lazarus Trial. The treatments would involve stem cell and amino acid injections into the brain, laser therapy and nerve stimulation, in an attempt to bring the central nervous system back to life. Several neurologists and neuroscientists believe that these techniques may one day be of value in reawakening comatose patients but that revitalising already clinically dead patients seems unlikely.

REGENERATION OF DISEASED ORGANS

As was mentioned at the beginning of this question, some creatures have the ability to generate new body parts if the originals become diseased or damaged. Starfish, for example, can re-grow missing arms. A relative of the salamander, the axolotl, can regenerate its tail, limbs, skin and almost any other body part. Researchers found that each time a limb was removed, it re-grew almost perfectly. This creature has become a focus for research into human tissue regeneration.

It seems that the power for complex animals to re-grow damaged body parts has been lost during evolution. If the mechanism can be unravelled and controlled, then perhaps it would become possible for us to re-grow ageing hearts, kidneys and blood vessels. That would certainly lead to an increased lifespan.

SPARE-PART SURGERY AND 3D PRINTING OF ORGANS

Organ failure, whether heart, kidney, lung, etc. is a common cause of death. Organ transplantation has now become almost routine, the main problem being a shortage of donors. Patients with end-stage heart, kidney or lung disease could have their lives extended if suitable organs were available. Recent advances in 3D printing technology could, in the foreseeable future, render the need for human donors obsolete. Replacing aged and non-functioning organs with new synthetic ones could become a viable option for life extension.

How much could human lifespans be increased by if one day a method for slowing or preventing ageing can be shown to be safe and to work? It is hard to say but it could conceivably be by several decades or even longer. We will have to leave aside the difficult questions of how to house and care for and fund the rapidly expanding and aging population that would result from this, when we already seem to have too many people in the world.

THIS IS AN UNANSWERABLE QUESTION

Best Guess Answer

No human has lived longer than 122 years, at least not with proper documentation to prove it (and even that is in some doubt). A healthy lifestyle will help more people to live longer but is unlikely to increase the current lifespan significantly. New technology would be needed for that to happen. Since ageing is a process, it should theoretically be possible to interfere with it and slow it down.

Possible routes to extending lifespan can be divided into the following categories:
a) Adopting a healthy lifestyle
b) Medication to interfere with and slow down the ageing process

c) Cryopreservation and re-animation when future treatments are available for the original cause of death
d) Spare-part surgery and 3D printing of organs

However, the social and economic problems created by a population containing 500-year-old multi-great grandfathers and grandmothers are likely to be insurmountable. Of course, not everyone would wish to live that long but some might. Perhaps there would be different strengths of tablets offering differing lifespans—100mg for every 100 years maybe?

Having said that, current work in this area is focused more on keeping people well into their old age rather than trying to make them live longer; a fit and healthy 85-year-old rather than a decrepit 105-year-old.

Our answer is that lifespan-enhancing procedures for a modest (a few decades) increase are feasible. More extensive increases in the 100-or-more-years range are theoretically possible but unlikely to become available for many decades, maybe even longer, if they are allowed at all. Living forever is almost certainly impossible.

Q9

HOW AND WHEN WILL THE WORLD END?

The end of the world is not something that most people think about since it is a depressing thought. If you are reading this then at least this prophecy (Image 111) was wrong, as have

Image 111. *South Bank, London, May 2018. ©Author.*

been all the others from cults, religious groups and historical figures over the years.

We can however approach this topic more scientifically. The world, that is the Earth and the rest of the universe, had a beginning as discussed in the first four questions so it is not unreasonable to suppose that they will also have an end. What form might this take and when might it happen?

Before embarking on this discussion however it should be made clear that everything that follows is based on the assumption that the laws of physics are the same everywhere in the universe and have been so since it began. There is no evidence that this is a false premise but that is not the same as saying it is true. It just has not been falsified yet. It is however a necessary assumption to make until we have evidence that it is incorrect.

Let us first list the possibilities regarding the Earth and the Sun. We will come to the rest of the universe later. These divide into man-made scenarios and natural scenarios.

MAN-MADE SCENARIOS
Global biological, chemical or nuclear wars

It is conceivable that a global-scale war involving the use of some or all of the above weapons could destroy all life on Earth. With current sophisticated monitoring facilities, however, it seems unlikely that one nation would be able to mount a surprise attack that would devastate the entire planet. In any case, such a result would hardly be of benefit to the aggressors.

So, while it obviously remains a possibility, the prospect of Man destroying all life on Earth through an act of war, or perhaps through a genuine mistake, seems unlikely. Our

Image 112. *Atomic bomb explosion. ©Wikimedia Commons.*

biggest threats actually come from a variety of possible natural occurrences.

NATURAL SCENARIOS

Astronomical impact

There is a large variety of objects moving around the Solar System apart from the Sun, its planets and moons. These range from minute dust particles to gigantic lumps of rock and our concern here is with those that could conceivably collide with the Earth, that is, asteroids, comets and meteors.

Asteroids are small rocky bodies that orbit the Sun between the orbits of Mars and Jupiter, as shown in Image 113. There are many millions of them ranging in size from hundreds of kilometres in diameter downwards.

There is a second band of asteroids outside the orbit of Neptune known as the Kuiper belt. Discovered in 1992, the Kuiper belt is similar to the asteroid belt, only much larger, that is, at least 20 times as wide. Pluto, reclassified as a dwarf planet in 2006, is a member of the Kuiper belt.

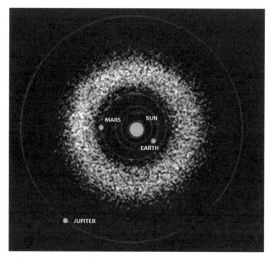

Image 113. Diagram of the Solar System showing the position of the asteroid belt. ©Wikipedia.

Comets are lumps of rock and ice that can be tens of kilometres wide. They are characterised by their long tails composed of ice and gas vapours produced as they pass close to the Sun. These tails can be enormous, up to 80-million-km (50-million-miles) long although the solid part of the comet, the nucleus, may only be a few kilometres long.

There are about 4,000 known comets but undoubtedly there are thousands if not millions of others that remain to be discovered. The most famous is that named after Edmund Halley, the British astronomer who first worked out, in 1705, that this known comet would return to be seen from Earth approximately every 75 years (Image 114).

Halley predicted a return in 1758 but died in 1742 so never lived to see his prediction come true. The most recent appearance was in 1986 as shown in Image 115.

Halley's Comet, although it was not called that then, was depicted on the Bayeux tapestry (Image 116).

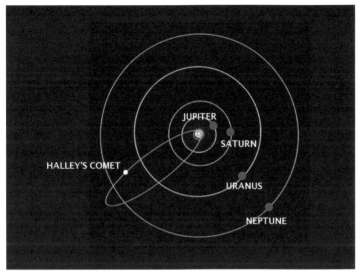

Image 114. Diagram showing the orbit of Halley's Comet in relation to the orbits of the outer planets. ©Wikipedia.

Image 115. Halley's Comet in 1986. ©Wikimedia Commons.

***Image 116**. Bayeux Tapestry with spectators pointing at Halley's Comet (top right) during its appearance in March 1066. ©Wikipedia.*

Interesting Aside: *No one knows for sure how Halley pronounced his surname. The most common version would probably be to rhyme with valley but Halleys in the US apparently prefer it to rhyme with daily. The rock singer Bill Haley called his backing group The Comets in deference to Edmund Halley.*

A meteor (also called a 'shooting star') is actually the streak of light that occurs when an object, that is, a meteoroid, enters the Earth's atmosphere. Most are about the size of a beach pebble and will usually disintegrate during their journey through the atmosphere. Those that do land intact are known as meteorites.

Asteroids, comets and meteorites can and do collide with each other, and with planets and moons. The familiar

***Image 117.** Leonid meteor shower 2009. ©Wikimedia Commons.*

cratered surfaces of our Moon and others in the Solar System are due to multiple collisions with such objects in the past. In 1994, Jupiter suffered an impact from the Shoemaker-Levy 9 comet which was observed by NASA spacecraft cameras.

The Earth has also had its fair share of impacts. Image 118 shows the Meteor Crater in Arizona probably formed by an impact with a 30m wide meteorite weighing over a quarter million tons. Unusually, this crater is privately owned.

One of the largest impact craters on Earth is the Chicxulub Crater buried underneath the Yucatán Peninsula in Mexico. It is more than 160km (100 miles) wide and was formed 65 million years ago. Although not proven, many scientists believe that this collision caused a series of mass extinctions of plants and animals including the dinosaurs.

A remarkable recent discovery is the Kamil Crater in the Egyptian desert. This is a 45m wide x 16m deep (150ft x 50

Mysteries of the Universe 199

Image 118. Meteor Crater, Arizona, USA. ©Wikimedia Commons.

ft) impact crater that was discovered in 2008 by an Italian scientist Vincenzo de Michele while looking through Google Earth satellite images.

Ground-based studies of the Kamil Crater in 2010 revealed that it was caused by an impact with an iron meteorite weighing about 5 to 10 tons with the impact occurring approximately 5,000 years ago. The meteorite is estimated to have been about 1–2m (3–6 ft) wide.

If an object, whether an asteroid, meteoroid or comet, is in an orbit that brings it close to the Earth, then it is known as a Near Earth Object. It is comforting to know that several nations are undertaking a continuous monitoring of these objects, especially those of a size sufficient to cause major damage on impact.

As might be expected, collisions with smaller objects are far more common than collisions with larger objects. This of course is just as well since the small objects do little or no damage whereas the really large ones can have devastating effects.

Here is a table showing the approximate relationship between size (in metres) and frequency of impact.

Table J. Frequency of impacts

DIAMETER	FREQUENCY
< 10 m	every year
10 – 50m	every 5 years
100 m	every 1,000 years
1 – 2 km	every ½ million years
15 km	every 50+ million years

Smaller objects would be likely to vaporise in the atmosphere and cause no effects. This occurred on 30 June 1908 near the Podkamennaya Tunguska river in Russia. The Tunguska Event as it is called, is believed to have been caused by the air burst of a meteoroid or comet at an altitude of 5 to 10 km (3 to 6 miles). The object is estimated to have had

Image 119. Flattened trees at Tunguska, 1908. ©Wikimedia Commons.

a diameter of tens of metres. There was no crater as there was no impact as such but the force of the air burst knocked down an estimated 80 million trees over an area greater than 2,000 square km (800 square miles).

These scenarios have been a favourite subject for science fiction writers and film producers. *When Worlds Collide* was a popular 1950s disaster film.

But, could the Earth actually be destroyed as a result of an astronomical impact?

Yes, it could, although it would be an extremely rare event (obviously, as it has not happened over the past 4.5 billion years). However, one could argue that since the last massive impact was 65 million years ago (although that did not actually destroy the Earth), perhaps we are due for another one soon. On the positive side, there are now many sophisticated surveillance devices both on Earth and in

Image 120. Book cover of When Worlds Collide, *1933. ©Frederick A. Stoker.*

spacecraft that are continuously monitoring the movements of numerous Near Earth Objects, especially the really large ones. Should such an object be detected and should its orbit bring it on a collision course with Earth, there would usually be many years of warning.

In April 2019, it was announced that a 1,200 ft (370 m) wide asteroid discovered in 2004 and known as Apophis was on a trajectory to pass within 19,000 miles (31,000 km) of Earth, on Friday 13 April 2029. That is extremely close but astronomers are certain that it will not collide with Earth, instead giving them an unprecedented opportunity to study it closely, and perhaps even sending probes to investigate it directly. This would provide important information and may help in future encounters when collisions could be likely.

Although it is theoretically possible, the chances of suddenly finding a massive Near Earth Object that is calculated to impact in hours or days is extremely remote.

The whole subject of Near Earth Objects and possible collisions with the Earth is now an entire topic of its own. Spaceguard is one of several international organisations whose task is to monitor the threat posed by Near Earth Objects. There is even a hazard scale, known as the Torino Scale, which runs from zero (no risk) up to ten (certain collision causing a global catastrophe). Methods for preventing an impending collision, probably by use of space-borne high-power lasers or nuclear explosions, are being considered and researched by the relevant authorities.

An astronomical impact with a sufficiently massive asteroid or comet could end all life on Earth. It is possible but extremely unlikely.

Finally, as discussed in Question 5, the probable collision of the primitive Earth with another large celestial body, eventually forming our Moon, was an impact that may actually have made life on Earth possible rather than destroying it.

GAMMA RAY BURST

A gamma ray burst is a flash of extremely powerful gamma rays emitted from a supernova (exploding star, Image 121). The first one was detected by a spy satellite in 1967. All the ones that have been observed so far have originated in distant galaxies billions of light years away.

The fact that the radiation has travelled so far and is still easily detectable on Earth gives some impression as to how powerful it really is.

It has been estimated that a single gamma ray burst releases as much energy in a few seconds as our Sun will do in 10 billion years. It has also been estimated that gamma ray bursts occur about once in every 100,000 to 1 million years per galaxy.

On 23 April 2009, the American Swift satellite detected a gamma ray burst (Image 122) which turned out to be the most distant, and hence the oldest, object known in the entire

Image 121. Diagrammatic representation of a gamma ray burst resulting from a supernova. ©Wikipedia.

Image 122. Gamma ray burst GBR 090423 (circled). ©NASA.

universe. It was measured as being 13 billion light years away. This means that the light and gamma radiation detected by the satellite started its journey 13 billion years ago. Since the universe is considered to be 13.7 billion years old, this object comes from an era when the universe was only about 700 million years old.

No gamma ray bursts have so far been detected as originating in our galaxy. That is just as well since it is possible that if that were to happen, and the burst was beamed towards Earth, it could end all life on the planet.

OTHER NATURAL EVENTS

Tsunamis, earthquakes, floods, volcanic eruptions, climate change, overpopulation, famine and disease can all be of huge proportions and cause massive destruction of property and life.

For example, the Spanish flu outbreak of 1918–20 was estimated to have killed up to 100 million people worldwide. However terrible, none of these events is likely to threaten the viability of life on Earth, or of the Earth itself.

LIFE CYCLE OF THE SUN

What we have considered so far—wars, astronomical impacts, gamma ray bursts and other natural events—may or may not happen. There is really no way of knowing. But, based on detailed studies of other stars, we do know that our Sun has a well-defined and finite life cycle, as shown in the diagram below.

As the Sun's store of hydrogen diminishes, so its temperature, and that on the Earth, will increase. In a billion years or so its temperature and consequently that on Earth will rise so much that liquid water can no longer exist. Even before that, the surface temperature is likely to be too high to support life and unless our descendants find a way around this problem, all life on Earth will become extinct.

As the Sun continues to become hotter over the next few billion years it will also get larger. In about 5.5 billion years it will become a red giant, many times larger than it is now. It will engulf Mercury and Venus and possibly Earth as well. It will then collapse into a white dwarf and remain as such for many more billions of years.

These timings are only approximations but the eventual outcome is pretty certain. Life on Earth, and Earth itself, will be destroyed.

That sounds bad. However, a consequence of the Sun getting hotter is that more heat is available to and will reach

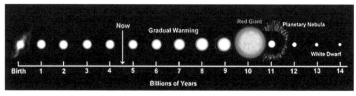

Image 123. Life cycle of the Sun. ©Wikipedia.

the outer planets and their moons. Some of these moons, such as Europa, a moon of Jupiter, and Titan, a moon of Saturn, amongst others, have been suggested as possible sites for the emergence of life.

It is conceivable therefore that in a few billion years or so, when the Sun is getting too hot for us here on Earth, the entire population will have migrated outwards to one or more of these moons. The extra heat given out by the Sun will make the surface temperature much more agreeable on these moons than it is now. That should give mankind a few more billion years of existence before the Sun finally becomes a white dwarf and, over many more billions of years, gets colder and eventually becomes unable to support any sort of life within its Solar System. By that time though, we should have worked out how to deal with this, probably by moving on to another Solar System.

THE FUTURE OF THE UNIVERSE

As discussed in Question 1, current thinking is that the universe began 13.7 billion years ago with what has been called the Big Bang. Rapid expansion during the first few fractions of a second (known as the period of inflation—see Question 1) and a subsequent slower expansion has resulted in the universe we see now.

One could argue that this presents three possibilities for the future of the universe:
1. It could continue to expand forever (the Open Universe model), also known as the Big Rip.
2. It could at some future time stop expanding and start to contract (the Closed Universe model), also known as the Big Crunch.
3. It could expand but at a slower and slower rate and never

contract (the Flat Universe model), also known as the Big Chill.

Can we differentiate between these three outcomes and predict which will happen?

The total amount of matter in the universe is the key. If this is below a certain amount, known as the critical density, the universe will continue to expand because there would be insufficient matter, and hence gravity, to halt the expansion.

Conversely, if the amount of matter is above the critical density, then gravity will eventually prevail and the expansion will slow down after which the universe will begin to contract, possibly resulting in a reverse Big Bang which has been called the Big Crunch. If that turns out to be the case, there could be a new Big Bang to start everything over again, and so on forever more. This is known as the Oscillating Universe model.

Here is a purely speculative scenario for such a situation.

With an Oscillating Universe model, we can make an enormous leap into a far-distant future, many quadrillions of years away. Given enough time, all random events will happen. Put another way, if something can happen, then given enough time, it will happen. All possible bridge hands will have been dealt; a coin will come up heads a million times in a row; all the molecules of air in a room will travel in the same direction at the same time and someone standing at the wrong end of the room will suffocate, etc. In his excellent book *One, Two, Three...Infinity* (1947), George Gamow gives a calculation for the waiting time for such a scenario in an average-sized room—$10^{299,999,999,999,999,999,999,999,998}$ seconds. To put this into perspective, the age of the universe is about 3×10^{17} seconds so there is no need to worry when you are in a room.

Gamow's book is still available and absolutely worth reading in spite of its age.

So, imagine this. Given enough time in an Oscillating Universe scenario, in our far-distant future, a universe just like ours will develop, containing a planet just like Earth, with life just like ours. One possible outcome out of the vast number of possible outcomes would be us and our families and friends.

With an Oscillating Universe therefore, we could all be reborn one day to live our lives again. It is an interesting, and perhaps to some, a comforting thought. And maybe our current lives are a repeat of an earlier one from a previous universe. We would, of course, have no memories of such putative earlier existences so this seems unlikely ever to be a provable hypothesis.

Returning now to the possible fate of the universe, we have seen that this is dependent on the amount of matter it contains. Cosmologists have calculated that the critical density (the amount of matter in a given volume of space) is equivalent to about ten hydrogen atoms per cubic metre. It turns out that the actual density of the universe is about the same as the critical density. This means that the universe contains just the right amount of matter eventually to halt its expansion—this is the Flat Universe model described earlier.

An astonishing discovery in 1998 threw all this into confusion. Astronomers working with the Hubble Space Telescope discovered that the universe was actually expanding and at an increasing rate.

Why was this astonishing? We can consider an everyday analogy by imagining a tennis ball being thrown up into the air. There is enough mass in the ball for gravity to act upon it and to slow its rise, stop it momentarily, and then drag it back to Earth. This is the Big Crunch model described earlier. However, the 1998 discovery would be equivalent to the tennis ball being thrown up into the air and, rather than slowing down and falling back to Earth, it carried on rising at a faster and faster rate.

The only explanation was that some unknown force was acting against gravity, a sort of negative gravity or repulsive force, pushing the galaxies further and further away from each other. This force has been called Dark Energy. If the expansion continues then the universe could end in a Big Rip. (See page 213 for more on Dark Matter and Dark Energy).

So, what can we say about the ultimate fate of the universe?

Up until the late 1990s, scientists had two models of how the universe's expansion would work. In one scenario, there would be enough matter in the universe to slow the expansion to the point where, like the ball, it would come to a halt and start to retract, everything crashing back together in a Big Crunch.

In the other scenario, there would be too little matter to stop the expansion and everything would drift on forever, always slowing down. The galaxies would drift apart from each other until they were out of view because they would eventually be so far away from us that the light by which we see them would not have time to reach us. The universe would continue growing larger as countless generations of stars faded and died out. It would end in a vast, dark and cold state, sometimes referred to as the Big Chill. This would be equivalent to the death of the universe. Why? Because of the Second Law of Thermodynamics.

Imagine a glass of water with some ice cubes in it. Now imagine the same glass 15 minutes later. What has happened?

Images 124 & 125. *Glass of water with ice cubes (left) and then 15 minutes later (right). ©Author.*

As shown by the photographs (absolutely genuine—it really is the same glass in the same position 15 minutes later), the ice has melted and the water level has dropped because ice takes up more volume than the water it is made of due to its crystalline structure.

The ice has melted because the temperature in the room is higher than the temperature of the ice and the ice has melted because heat always flows from a warmer to a colder place and never the other way around. A hot body has more energy than a cold one, and energy always flows so that the energy—temperature in this case—becomes equalised.

Eventually, the temperature of the water in the glass will be the same as that of the air in the room which will have become a little bit colder due to the energy taken out of it to warm up the ice and water. The energy between the room and the water will then have become equalised.

In simple terms, this is the Second Law of Thermodynamics—left to its own devices, heat always flows from a warmer to a colder place until the temperatures have been equalised.

This is related to the concept of entropy. Entropy is an expression of the state of disorder of a system. The greater the disorder, the higher the entropy and vice versa. Nature always tends towards a state of maximum entropy for any system left to its own devices. This is a scientific way of saying that the natural tendency is for things to go from order to disorder. Ice had an ordered crystalline structure whereas liquid water does not.

If you carefully build a house of cards the chances are that it will have collapsed after a few seconds as shown in Images 126 and 127 (it actually took about 2 seconds). Its disorder, or entropy, has increased.

So, in the Big Chill model, given enough time, the universe will reach a state of maximum entropy, that is, all the energy in the universe will have been equally distributed and there will be the same temperature and pressure everywhere and everything will be randomly arranged with no structure. This means that no chemical or physical reactions can take place, nothing happens and everything gradually stops. The universe has died.

Images 126 & 127. *Card house and its collapsed condition after a few seconds.* ©*Author.*

But do not worry too much. This state of affairs will be a very very long time coming if it comes at all.

By the early 1990s, astronomers had calculated how much mass was in the universe and decided on the Big Chill as the most likely end of the universe. According to the Big Chill, the universe should be expanding more slowly today than it did in the past because gravity has had time to work on slowing the universe down over all these billions of years. But as described above, astronomers found that the universe is actually expanding, and more rapidly than it was in the past, meaning something must be working to speed it up.

This result seemed anomalous because gravity always pulls and slows—it never pushes. Yet some force did appear to be pushing the universe apart. Not knowing what this force was, it was called Dark Energy.

With Dark Energy, the fate of the universe might go well beyond the Big Chill. In the strangest and most speculative scenario, as the universe expands ever faster, all of gravity's work will be undone. Clusters of galaxies will disband and separate. Then galaxies themselves will be torn apart. The solar system, stars, planets, and even molecules and atoms could be shredded by the ever-faster expansion. The universe that was born in a violent expansion could end in another, called the Big Rip.

So out of the three scenarios for the fate of the universe—collapse to a Big Crunch, expand ever more slowly to a Big Chill or expand ever faster to a Big Rip, it has been possible to narrow the possibilities.

Evidence has ruled out the Big Crunch. The Big Chill is a possibility. Whether or not the universe goes all the way to a Big Rip depends on what Dark Energy really is and whether it will stay constant forever or fade away as suddenly as it appeared. And that we do not know yet.

Here is a brief summary of the Big Crunch, the Big Chill and the Big Rip.

Big Crunch

The expansion of the universe stops and gravity makes everything come together again in a reverse of the Big Bang.

Big Chill

The expansion of the universe eventually slows down when the energy of the universe is equally distributed everywhere. Everything has the same temperature and pressure, no chemical reactions take place, and everything has stopped forever.

Big Rip

The expansion of the universe gets faster and faster and eventually everything, even atoms, gets ripped apart leaving just sub-atomic particles floating about.

No matter which scenario is most likely, the universe still has at least a few tens of billions of years left, which leaves us plenty of time to look for the answers and solutions.

DARK MATTER AND DARK ENERGY

Dark Matter

Planets revolve around the Sun at speeds dependant on their average distance from the Sun—the greater the distance, the lower the speed. For example, Mercury, the closest planet to the Sun, revolves at a little over 100,000 mph (165,000 kph), Earth revolves at 66,000 mph (105,000 kph), and Neptune, the furthest planet from the Sun, revolves at just 11,000 mph (18,000 kph).

This relationship is governed by Kepler's Third Law of Planetary Motion (see Appendix B). However, these details do not matter for now; what is important is the principle that in a revolving system like a solar system or a planet with moons, where most of the mass is in the centre, the further out an object is from the centre, the slower it will revolve around the centre. In other words, the rotation speed declines with distance.

In the early 1930s, during studies of revolving galaxies, it was found that stars revolve around their galaxy's centre at equal or even increasing speeds as their distance from the centre increases. This was a totally unexpected finding which went against known physical principles.

Let us try and put this into perspective with an everyday example. Consider a see-saw. This will be balanced if there is equal weight on both sides, and if each weight is at the same distance from the pivot point, which is the situation in Image 128.

Image 129 shows the see-saw with twice the weight on the right-hand side at half the distance from the pivot point, in which case the see-saw will still be balanced. This is because physics tells us that as long as the weight multiplied by the distance from the pivot is the same for both sides, there will be a balance.

Image 128. *Equal weights at equal distances.* ©*Author.*

***Image 129**. Twice the weight at half the distance. ©Author.*

Now look at Image 130. Here we have equal weights at equal distances, as in the first example, so there should be a balance. But there is not. The left-hand side is lower, indicating that this side is heavier than the other one even though we know that both weights are the same and both are at the same distance from the pivot point.

So how can we explain this anomaly? One explanation is that the left-hand side of the see-saw is somehow heavier than it appears even though we know that both weights are the same and that they are at the same distance from the pivot point. Somewhere on the left-hand side there is extra weight we do not know about and cannot detect.

In the same way, astronomers struggled to explain the anomalous rotation speeds of stars in spiral galaxies. They just did not behave as if most of the mass of the galaxy was concentrated in its centre. One way to explain this was to surmise that there must be more matter in the galaxy than

***Image 130**. Equal weights at equal distances. ©Author.*

they could detect, and that this matter was more evenly distributed throughout the galaxy than was apparent from visual examination. This would then explain the faster-than-expected rotation speeds of the outer stars in the galaxy. Since this extra matter was undetectable, it was called Dark Matter. Current estimates are that Dark Matter makes up nearly 25 percent of all the matter in the universe (Image 131).

It should be emphasised however that the concept of Dark Matter is just that—a concept. Its existence has not been proven and it may turn out that there are other explanations for the anomalous behaviour of rotating galaxies.

Dark Energy

Dark Energy is another hypothetical entity proposed to explain another anomaly—the observed expansion of the universe as discussed on page 208. In 1998, it was found that the universe is actually expanding and at an increasing rate. The anomaly arises because the amount of matter in

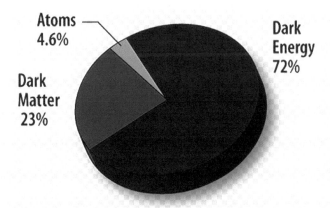

Image 131. Estimated amounts of Dark Energy, Dark Matter and ordinary matter (atoms) in the universe. ©NASA/WMAP.

the universe, even including the hypothetical Dark Matter, is such that the expansion of the universe should be slowing down due to gravitational effects, not speeding up.

So how can this anomalous finding be explained? Enter Dark Energy, another hypothetical entity that is theorised to occupy all of space and act by pushing galaxies apart, a sort of repulsive gravity which explains their increased rate of expansion.

Calculations show that Dark Energy can account for over two-thirds of all the matter (energy and matter being related as per Einstein's equation $E = mc^2$) in the universe.

It should be emphasised that neither Dark Matter nor Dark Energy have ever been directly detected. They are both theoretical inventions created to explain astronomical observations. They may or may not turn out to be correct.

Such theoretical concepts are common throughout science and serve to offer explanations for otherwise unexplainable observations. (Look up Phlogiston Theory of combustion; epicycles and planetary motion; and the aether and the transmission of light. These were all theories to explain observations; all of them turned out to be wrong).

THIS IS AN UNANSWERABLE QUESTION

Best Guess Answer

Life on Earth and Earth itself face realistic threats to their existence. Life could be seriously compromised and even annihilated as a result of nuclear, chemical or biological wars. We have to hope though that no one would be that stupid. So far, even though these weapons have been available for decades, we have escaped, and we have to hope that our governments will continue to keep us safe.

Natural climatic and infectious disasters also pose a threat, but none are likely to be of sufficient magnitude to actually wipe out all life on Earth or destroy the Earth itself.

There could however be truly devastating consequences from an astronomical impact or from a gamma ray burst. Fortunately, these are extremely rare events which are closely monitored and though it is possible that we could succumb to such a thing, one has to say that it is very unlikely.

In time though, we will all succumb to the laws of physics. Life on Earth, and Earth itself, will eventually come to an end when our star, the Sun, enters the later stages of its life cycle.

These events are billions of years into the future and by then we should have developed ways of emigrating to one or more of Jupiter and/or Saturn's moons, which could provide a home for a few more billion years. After that, when our Sun is on the way to becoming a white dwarf, we would have to go a bit further afield to a new solar system.

That seems like quite a journey for a few billion people but in a few billion years, who knows what we will be able to do?

Looking for and moving to a new solar system as a home for all life on Earth seems a daunting task but dwindles into insignificance when compared to finding a new universe. If ours continues to expand, distant future generations may have to deal with either a Big Chill or even a Big Rip. Their best hope then is that we do indeed exist within a multiverse (Question 2) and that they can find a way to get to another life-sustaining universe.

Q10

IS THE FUTURE PREDETERMINED?

"Nothing can be said to be certain except death and taxes."

So said Benjamin Franklin in 1789. While it is certainly true that some future events are inevitable, almost all of them are a consequence of the laws of nature, and so would include predicted future eclipses, the decay of radioactive materials, the motions of the planets and stars, etc. To this list could be added the eventual death of all living creatures. However, when people speak of the future being predetermined or preordained, they are referring to personal life events which is something completely different.

Those who subscribe to the notion of a predetermined future believe that we are powerless to do anything except what we actually do. Put another way, we have no power to influence future events or even our own actions. This philosophy is an integral part of various religious beliefs.

Let us imagine that you are unable to leave home to meet a friend because your car has a flat tyre. You go back inside and phone to apologise and then notice that your kitchen tap is leaking.

Had you left home as planned, you could have returned home to a flooded kitchen. For a believer, the flat tyre was

Images 132 & 133. *Flat tyre; leaking tap. ©Author.*

preordained so that the water leak could be dealt with before it caused significant damage. For a non-believer, the two incidents are unconnected.

What if it had been the other way around? You are about to leave home for your meeting when you notice that the kitchen tap is leaking. You phone to apologise and deal with the leak.

Once this has been done you go to your car and see the flat tyre. Instead of thinking that the flat tyre had been preordained so that you could prevent your kitchen from flooding, it has now become yet one more disaster to disrupt your arrangements.

So, flat tyre followed by leak equals a good thing, and leak followed by flat tyre equals a bad thing.

When you eventually get to meet your friend, what will you say?

Version 1: "Sorry I'm so late but I had a flat tyre. Really lucky it happened when it did because when I went back inside I saw that the kitchen tap was leaking. Could have been a disaster. Someone's looking after me!"

Version 2: "Sorry I'm so late but the kitchen tap was leaking. And then to make things even worse I had a flat tyre. Not my day!"

So here we have two completely unrelated events where the interpretation depended entirely upon the order in which they were noticed. For a believer, Version 1 will be retold as evidence of a preordained future and Version 2 will be retold as a bad day. In other words, a believer in a preordained future is likely to count only the positive outcomes.

Two or more things happening together is what is usually called a coincidence. A coincidence can be defined as an accidental occurrence of events at the same time with no causal relationship.

An example might be the chance meeting of a friend in a cinema or a sudden phone call from a relation living abroad who you had just been thinking about.

The important part of the definition is that there is no causal relationship, that is, the events happened independently by chance—one did not cause the other.

Image 134. *Surprise phone call. ©Author.*

Nevertheless, many people attach great significance to coincidences believing that they were meant to be or ordained or the result of divine intervention. This is because some coincidences can indeed seem remarkable and far beyond the powers of mere chance. In most cases though this is an illusion; many things that seem unlikely are actually much more likely than people might imagine. Many people also misinterpret or do not fully understand the power of chance.

Million-to-one events happen frequently as long as enough people take part. Look at the UK national lottery where 45 million-to-1 jackpots are won regularly because tens of millions of people buy tickets.

Let us have a look at a few examples of apparently spectacular coincidences.

THE BIRTHDAY SURPRISE

Ignoring leap years, how many people would there have to be in a room to be certain that at least two of them share a birthday? Since there are 365 possible birthdays, the answer is 366 people.

But how many people would be needed for the chance of coincident birthdays to be better than 50 percent? The answer will surprise you—it is just 23 people! It is true; with 23 people there is a better than even chance that two of them will share a birthday. With 57 people, there is a 99 percent chance that this will happen.

The explanation requires some understanding of probability theory and is well documented in books and on the internet for those who wish to explore this further. It is a very good example of an event that seems remarkable but is not really.

Another unexpected statistical result arises if a group of 30 people are each asked to write down any number from 1 to 100. What are the chances that two people will write the same number? Again, the result may surprise you—with 30 people it is over 80 percent!

DREAMS THAT COME TRUE

Sometimes someone has a dream which then apparently comes true. For the dreamer, this can be a remarkable experience. Say, for example, you have a dream about someone in a supermarket who falls over while carrying bags of shopping.

Then a few days later, you are in a supermarket and someone falls while carrying bags of shopping. What do you make of this? You will tell lots of people that you had a dream about someone in a supermarket who fell over while carrying bags of shopping, and it came true.

So, what does this mean? Did your dream foretell a preordained future or was it just a coincidence? Quite reasonably, you might think that it was such a remarkable coincidence that maybe it did mean something.

Most people dream about five times a night every night even though they may not always remember their dreams.

With 60 million people in the UK, that is 300 million dreams every night and over 100 billion every year in this country alone. In the US and Canada, it amounts to over 600 billion dreams every year, and in the Indian sub-continent there would be over 3 trillion dreams every year. That is an awful lot of dreams and some of them will bear some resemblance to future events.

What needs to be remembered are all the many billions of dreams that do not come true. Someone somewhere someday will have a dream that comes true. It does not mean anything; it is just a statistical result from a vast number of dreams. And of course, it is only that dream that will be retold.

THE WRECK OF THE TITAN

This is the title of a book that was written in 1898 by Morgan Robertson (it was originally called *Futility*). It tells the story of a disgraced former Navy man who worked as a deckhand on a fictitious ocean liner called the Titan which then hit an iceberg and sank.

In the novel, the Titan hit an iceberg in the North Atlantic and sank with the loss of over 1,000 passengers. Bearing in mind that the book was written in 1898, 14 years before the Titanic had its maiden voyage and even before it had been designed, the similarities between fact and fiction were indeed remarkable.

Wikipedia highlights some of the similarities as follows:
- Both ships had similar names;
- the Titan was 800-ft long and displaced 45,000 tons of water;
- the Titantic was 882-ft long and displaced 46,000 tons of water;
- both ships were described as 'unsinkable';
- both ships had a shortage of lifeboats;
- the Titan, moving at 25 knots, struck an iceberg on its starboard side on a night in April in the North Atlantic, 400 nautical miles from Newfoundland;
- the Titanic, moving at 22.5 knots, struck an iceberg on its starboard side on the night of 14 April 1912 in the North Atlantic, 400 nautical miles from Newfoundland;
- the Titan sank and 2,487 of its 2,500 passengers and crew died;

- the Titanic sank and 1,523 of its 2,200 passengers and crew died.

So, what did this mean? Did Morgan Robertson have the ability to foretell a preordained future? It is unlikely. He was a merchant seaman so he knew about ships and maritime matters. Icebergs were a known hazard in the North Atlantic; between 1882 and 1890, 14 ships were lost due to collisions with icebergs. So he conceived a story about a disgraced naval officer who makes good with a back story about a large ocean liner that sinks after striking an iceberg in the North Atlantic.

As shown above, there were indeed many similarities between the story and the real event but there were many differences as well. If it really was more than just coincidence, then we might expect all of the details to have been correct. Also, Robertson himself, after being told that he must be clairvoyant, brushed this off by saying that the similarities were due to his extensive knowledge of the subject.

The text of the book is available as a free e-book download from Amazon.

CHANCE MEETINGS

We sometimes meet someone we know unexpectedly and an interesting or perhaps useful conversation might result. Was the chance meeting preordained?

How many people do you know? For most of us, the answer is probably in the mid- to high hundreds if we include family, friends, neighbours, acquaintances, work colleagues and trades people. You can get an approximate idea as follows.

Take a reasonably common name such as Michael. Several internet sources will be able to provide data on how many people have this name in the UK or in any other country; one such site is www.kganswers.co.uk.

The number provided for Michael is 673,000 (to the nearest thousand). The population of the UK when the data was researched (2007) was about 61,000,000. This means that 1.1 percent of the population was called Michael.

Now work out how many Michaels you know personally. They do not have to be close friends or family; you just have to know them even if you have not seen them for five years (but they do have to be alive). I could list 13 Michaels. Assuming that the percentage of Michaels in the population is the same as for my personal sample, then this means that 13 represents 1.1 percent of all the people I know. This calculates as close to 1,200 people.

You can increase the accuracy of the estimate by repeating the exercise with other names.

How many places do you go to and travel through during the course of a year? It is hard to give an accurate answer but if on average we are in five to ten different locations each day (bus, train, shop, road, work, restaurant, cinema, club, library, airport, school, etc.), it would add up to about 2,000 places in a year.

Let us err on the side of caution and say that most of us know about 750 people and that, during the course of a year, we spend at least some time in 2,000 different places. That amounts to 750 x 2,000 = 1.5 million possibilities of unexpectedly meeting someone we know per year. Perhaps not such a chance meeting after all.

Do not forget, we are talking about an event after it has happened. "That was quite a coincidence, meeting Alex in the coffee shop." That is a completely different scenario from saying "I wonder if we'll meet Alex in the coffee shop." The first sentence would have been a valid response if the meeting had been with any of the estimated 750 people that you know, whereas the second sentence would only have been valid if the named person was the one you met.

THE MEISSEN PLATE

Here is something that happened to me. My parents had a set of Meissen porcelain which had been in the family for many years and of which they were very fond.

Shortly after my mother died my wife and I were browsing around a local antique fair, as we did quite often, when we came across a stall selling a Meissen plate of the same pattern as our family set.

This was the first time I had ever seen one of these plates at an antique fair. Meissen have a vast number of different designs and to find one of the same rose pattern design as the rest of our set, and so soon after my mother had died, was quite a coincidence.

Image 135. Rose pattern Meissen plate. ©Author.

I bought the plate. It would have been very easy to think that finding this particular plate so soon after my mother died had some sort of significance—a preordained finding acting as a message perhaps to say that she was thinking of us and that we should keep and look after these family heirlooms.

Maybe. On the other hand, it could just have been a coincidence. Antique fairs are very popular with traders dealing in collectable porcelain and Meissen fits into this category.

I suppose it comes down to what you wish to believe. There is no 'evidence' here one way or the other but I prefer to follow William of Ockham's view that the most likely explanation is probably the right one.

TOTAL ECLIPSE OF THE SUN

The Sun has a diameter of 864,000 miles and is about 93 million miles from the Earth. The Moon has a diameter of 2,160 miles and is about 240,000 miles from the Earth. This means that the Sun is 400 times larger than the Moon and that it is on average very nearly 400 times further away. Because of this coincidence, when we look at the Sun and the Moon from Earth, both appear to be the same size. That is why we get total eclipses of the Sun when the Moon completely blocks out the Sun.

If the Earth to Moon distance had been, say, 1 million miles instead of 240,000 miles, that is, about four times further away, the Moon would look about four times smaller. Look what happens to the appearance of the total eclipse under these conditions.

Image 136 shows a total eclipse of the Sun as seen from Earth. The Moon completely obscures the Sun because

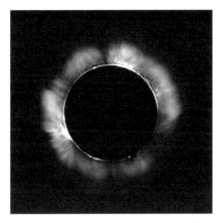

Image 136. Eclipse seen from Earth. ©NASA.

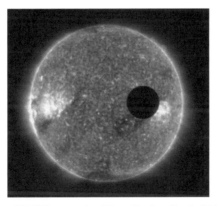

Image 137. Eclipse seen 1 million miles from Earth. ©NASA.

although it is 400 times smaller, the Sun is 400 times further away, so both appear the same size when viewed from Earth.

Image 137 shows the appearance from NASA's Stereo B spacecraft which photographed this passage of the Moon across the face of the Sun in 2007. Stereo B is in an orbit about 1 million miles behind Earth and was 4.4 times further away from the Moon as we are on the Earth's surface.

As explained here, this makes the Moon appear 4.4 times smaller when viewed from the spacecraft. So, we do not get a total eclipse from this viewpoint because the Moon appears too small. The apparent size of the Sun does not change significantly because it is already 93 million miles away and another 1 million miles makes little difference.

It is quite a coincidence that the apparent sizes of the Sun and the Moon as viewed from Earth are practically the same, making it possible to have a total eclipse of the Sun. No other Sun, planet, moon arrangement in the Solar System produces a total solar eclipse.

Is there any significance in this? Is it a preordained event to enable us to see a total eclipse of the Sun? No. It is just a coincidence. Bear in mind that the distance of the Moon from the Earth is slowly increasing by 3 to 4 centimetres per year.

That does not sound like much but eventually, in perhaps 1 billion years or so, the Moon will have moved far enough away so that its size as viewed from Earth will no longer be large enough to cover the Sun. At that time, total solar eclipses will forever be impossible.

THIS IS AN ANSWERABLE QUESTION

Answer

Coincidences are events that are time-related but not cause-related. They happen independently and even though some may appear remarkable, there is often a statistical reason for the concordance which may not be apparent to a non-mathematician. Other seemingly remarkable coincidences which cannot be explained mathematically are far more likely to be the result of pure chance than any other more esoteric reason.

When we come to consider whether life events are predetermined or preordained, we must resist the temptation to consider only those that support this notion. The flat tyre/kitchen tap leak scenario at the beginning of this chapter is a good example. We cannot legitimately select that order of events (tyre > leak) that results in a good outcome as evidence of a preordained future. If we adopt that type of investigative approach—only counting those results which support our thesis—then we could prove anything. The fact that both events happened at the same time is just a coincidence and the eventual outcome is purely dependent on which event is noticed first.

It is true however, that there are some situations where the feeling of a preordained future is compelling for the people concerned. Mr. and Mrs. Jones (names changed) were due to fly on Malaysian Airlines flight MH17 which was shot down in the Ukraine in 2014 with the loss of all 295 people on board, but were denied boarding because the flight was full. Mrs. Jones said "There must have been someone watching over us." The implication is that the 'someone' knew the flight would end in disaster, and so the crash was preordained.

Bearing in mind that this is a science book, we must restrict ourselves to scientific principles and ask for good evidence before making judgements. Although Mrs. Jones's response is completely understandable, it is not scientific evidence. What about all the other passengers who were not denied boarding? Here again we are cherry-picking just those outcomes that support the case.

It may be a comforting thought for many people to imagine that there are no real decisions for us to take in life since everything is preordained and nothing that we do will affect the future. It may also be a good way of coming to terms with disappointing and tragic outcomes—'It just wasn't to be'. However, none of these count as scientific evidence.

Apart from inevitable events resulting from the laws of physics and chemistry, nothing in life is predetermined, preordained or prearranged and we all have a completely free will to do and act as we please. Selecting just a particular sequence of events that has resulted in a positive outcome proves nothing, and coincidences, even apparently spectacular ones, do happen.

Q11

DOES PRAYER WORK?

Prayer is a religious practice that seeks to activate some form of connection to a God. It may be individual or communal and may take place in public or in private.

A prayer is likely to be offered for one of two reasons—to give thanks for good things or to ask for help for bad things.

Image 138. *Congregants at prayer in a cathedral.* ©*Author.*

The word is thought to come from the Old French word *preier* meaning 'to ask'.

Although different religions will have their own customs, rituals and forms of prayer, the basic idea is common. The big question is, does it work?

There is no doubt that things other than conventional medicine can make ill people better. When pharmaceutical companies test new drugs, they always run what are known as double-blind randomised controlled clinical trials.

Let us say we want to test whether a new drug is effective for treating high blood pressure. We would find say 400 people of similar ages and backgrounds with high blood pressure and divide them into two groups of 200 each. One group would be given the new drug and the other would be given a dummy pill (placebo) that did not contain any drugs but looked the same. The patients would be split up randomly and would be given a pill randomly. That means that the patients would not know whether they had the real pill or the dummy pill (in fact, they would only be told that they were taking part in the trial of a new drug) and the doctors running the trial would not know either. The trial runs for three months.

At the end of the three months, every patient is asked to confirm that they took their pills as required, and their blood pressure is measured and recorded. Only then are the doctors running the trial told which patients had which pills.

Here is what the results might have looked like.

Table K. Drug and placebo treatments

TREATMENT	PATIENTS	IMPROVEMENT	NO IMPROVEMENT
Dummy pill	200	40	160
Real pill	200	150	50

There is no chance of a bias here because no one involved in the trial knew who was taking real pills or dummy pills. The results are impressive. Seventy five percent of patients taking the new pill had a reduced blood pressure at the end of the trial whereas only 20 percent of patients taking the dummy pill had a reduced blood pressure. So, the new pill works.

But there is something else, perhaps even more remarkable. We said that only 20 percent of patients taking the dummy pill had a reduced blood pressure. But as it was a dummy pill, why should any of them have had a reduced blood pressure at all? After all, they had not been given anything apart from a dummy pill. It is called the placebo effect and is a well-known medical curiosity which first came to light during the Second World War when a doctor ran out of morphine and his nurse injected the patient with salt solution pretending that it was morphine. The operation was successful with the patient not suffering any discomfort.

Placebos can be effective in many conditions and can last for a considerable time. It probably works through substances known as endorphins which are produced by the body under certain conditions. These are similar to morphine and other opiates, and are pain relievers and also induce a feeling of well-being.

For our purposes though, it serves to illustrate that people can be made to feel better by giving them nothing. It is the belief that they have been given something that probably releases endorphins and these make the body feel better. Similarly, it is undoubtedly true that many people do feel better after prayer. Their belief that they have communicated with God does the work and again, endorphins may be the mechanism.

At that level, prayer certainly does work although probably not for the reason that the worshiper believes. But what about at a more serious level?

For example, what about the 800,000 Rwandan Tutsis who were killed during the genocide in 1994 and the many children who were orphaned? Did they pray to their Catholic God for help? You bet they did. Where was He when His people needed Him?

There are of course many more examples like these where unfortunate people are overcome by such horror that the placebo effect is powerless.

And in any case, even the strongest placebo effect would be useless when confronted with a ruthless enemy rather than with an illness. All the morphine-like substances in the world cannot protect you from machine guns or poison gas.

Those who pray also do it on behalf of others (intercessory prayer), usually in the hope that this might influence the outcome of a serious illness or a major operation or a dangerous mission. It is reasonable to assume that the person doing the praying may well feel better in the belief that they have done all that they could. However, there have been some properly controlled clinical trials on the effects of intercessory prayer (see summary for a reference) and no significant effects were noticed. If anything, those that received the prayers had a slightly worse outcome.

Approximately 6 million British soldiers enlisted in the First World War. It is a reasonable assumption that the majority of their families would have prayed for their sons' safety. In the event, about 700,000 were killed during the conflict. That is a percentage of just over 11 percent. What do we deduce from this? That prayer is 89 percent effective?

Those whose sons survived would say so. But ask those whose children died in the War and they would say that prayer was 0 percent effective.

It is far more likely that 11 percent represents the likelihood of being killed in such a conflict and that prayer has nothing to do with the survival rate. In any case, if prayer does work, why would the pleas of 700,000 people be ignored while protecting the remaining 5,300,000? There is just no sense in that.

Here is a true story. A local priest, the Reverend Neil Smith (not his real name) suffered a serious heart attack and was hospitalised for several weeks during which time he nearly died. It is fair to assume that, as a religious man, he and his family would have prayed for a good outcome. In this case, he did make a full recovery and returned to work.

I asked him whether he blamed God for his illness or thanked God for his recovery. He answered that God had better things to do than to strike him down with a heart attack for something that he may have done and that his illness was a medical matter and had nothing to do with God. However, he thanked God for his good recovery.

It is like the 33 trapped Chilean miners rescued from their mine in the Atacama Desert in 2010. These are deeply religious people who clearly would have prayed for their safe return but would not have blamed God for trapping them in the first place.

However, you cannot have it both ways. Either God is responsible for what happens or He is not.

What do religious leaders have to say about why bad things happen to blameless people? Over Easter 2011, the Pope, leader of over 1 billion Catholics, agreed to answer questions from people around the world. He responded to

7 out of more than 2,000 that were submitted. The first was from a young Japanese girl who said that she was very afraid after the earthquake and tsunami which killed many of her friends and classmates, and asked why this happened.

The Pope replied that he did not know why it happened but reminded her that Jesus also suffered and that he wants to help her with his prayers and that she could be sure that God would help her as well.

Why would God help this little child if he caused her misery in the first place? A more helpful and honest answer would have been to explain that natural events happen in unpredictable ways. No one makes them happen; they just do because that is how nature works.

THIS IS AN ANSWERABLE QUESTION

Answer

People who believe in a God probably feel better after prayer since their belief that they have communicated with Him releases endorphins from their pituitary glands and hypothalamus into the bloodstream, and this promotes pain relief and a feeling of well-being. At this level, prayer will work for some people. At a more serious level where the threat is from a powerful and ruthless enemy (such as war, personal attack, life-threatening illness), prayer is totally ineffective in influencing the outcome.

Religious people tend to thank God for good things but do not blame him for bad things. That is like saying that we will discard all the results of an experiment that do not give us the result that we want and only consider the results that do give us the result that we want. With that attitude you could prove anything. Ask Reverend Smith, or the Chilean miners, if prayer works. Of course, it does, and we are here to prove it. Ask them why they got into trouble in the first place and you will not get a straight answer, not even from as holy a person as the Pope.

There have been several clinical studies seeking to determine whether prayer is effective. *American Heart Journal* published a study on intercessory prayer (praying for someone else) for patients who had undergone heart bypass surgery (see *Am. Heart J.*, 2006, 151: 934–42 for the full paper). It found no significant results although those patients who received prayer actually had slightly worse outcomes.

Q12

ASTROLOGY: SENSE OR NONSENSE?

ORIGINS

Astrology has a long history, probably arising during the ancient Babylonian civilisations in about 1,500 BCE. Its basic tenet is that the relative positions and movements of celestial bodies, mainly stars and planets, can directly influence life on Earth and can also be used to provide information about an individual's personality and his or her present and future circumstances. This was perhaps an understandable view to take at the time since it was clear to ancient astronomers that the Sun and the Moon, which caused the tides, the seasons, day and night, for example, did indeed have an effect on life on Earth. Since these two celestial bodies could exert an influence, it was not much of a leap to suggest that other objects in the sky—the planets and stars—could also affect life on Earth.

These beliefs resulted in many cultures developing elaborate systems for predicting not only seasonal changes, weather patterns and eclipses, but also an individual's personal characteristics and future events. This was perhaps more of a leap than justified by observations but it then resulted in the more modern concepts of star signs and horoscopes.

SENSE OR NONSENSE

Whereas in reality the Earth moves around the Sun, seen from Earth it appears that the Sun moves around the Earth and this was the reason for the notion that the Earth is at the centre of everything.

This apparent path of the Sun, known as the ecliptic, is shown as a thick white line (P1) in the diagram. Astrologically, it is divided into 12 regions which together make up the Zodiac. Various groups of stars (constellations) lie within each of these regions. A constellation is a group of stars, apparently close together, that can be connected by imaginary

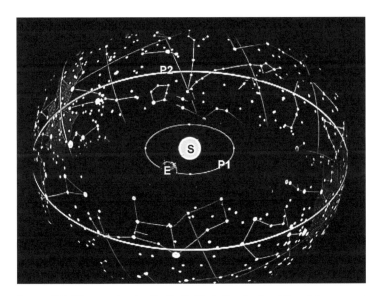

Image 139. Diagram showing the Earth (E) circling the Sun (S) and the apparent path of the Sun as seen from Earth during the course of one year (thick white oval line P1). The actual path of the Earth round the Sun is shown as P2. Modified from Wikimedia Commons.

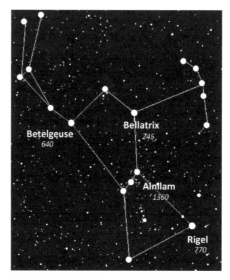

Image 140. The constellation of Orion (the hunter) with some of the stars joined up to form the shape of a figure wearing a belt and firing a bow and arrow. There are four named stars and their distances in light years from Earth. Modified from Wikipedia.

lines to form a recognisable shape, often an animal. (The word 'zodiac' derives from the Greek word for 'small animal'). Examples include Leo the lion and Pisces the fish. A non-animal example is Orion which can be joined up to appear like a human hunting figure wearing a belt and firing a bow and arrow as shown in the diagram.

Astrologers will determine an individual's star sign by consulting charts which show where the Sun is located on the ecliptic, and hence in which constellation, at any date in the year. It is then a simple matter to state that if, for example, an individual's date of birth is 10 October, then his or her star sign would be Libra since that is where the Sun would be located on that date.

A horoscope (from the Greek for 'observer of the hour [of birth]') is a generalised personality statement for people born under each of the 12 star signs, as well as a prediction for the day, week or year.

It is clear that there are many stars within a constellation's field that have been ignored when constructing the drawing. By selecting different stars, it would be easy to join them up to form any other shape—a house or a telephone or an aeroplane for example. Just selecting those stars that happen to form one particular recognisable pattern is completely arbitrary and meaningless (see the discussion of the Betty Hill star map in Question 5).

It is also important to appreciate that although impressive patterns may be seen from the viewpoint of someone on Earth, such as the photograph of Orion (and the joined-up stars to form the hunter and his bow and arrow), this is a completely false view of the actual situation.

The selected stars that make up the pattern have no relationship with each other and are separated by vast distances in all directions. Of the four named stars in the Orion photograph, Bellatrix is closest to Earth at 245 light years whereas Alnilam (the middle star of Orion's belt) is furthest away at 1,360 light years. These two stars are therefore separated by over 1,100 light years (about 7,000 million million miles), not in the plane of the photograph but in depth. This is suitably illustrated in Image 141 where plane A shows the apparent positions of the stars as a two-dimensional image viewed from Earth whereas the three-dimensional projection shows the reality.

Ancient astronomers noted that of the various objects in the sky, some moved across the sky and some were fixed. The moving objects were called planets (from the Greek word meaning 'wanderer') whereas the non-moving objects were called fixed stars (although these stars appear stationary

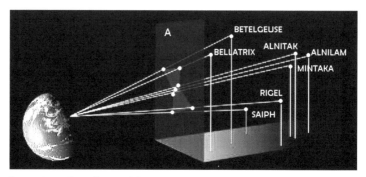

Image 141. Constellation of Orion as a 2D and a 3D projection. Modified from Wikipedia.

relative to each other, they do in fact move, but too slowly to be noticed by the unaided eye). Since the stars were supposedly fixed, it was assumed that they were embedded on the surface of a single celestial sphere, and the apparent view as in plane A was considered to be the reality.

To be fair, since this is exactly what it looks like when a constellation of stars is viewed from Earth, we can forgive the ancient Greek and other astronomers for their incorrect interpretation. Now though, we know better.

The apparent shape that we can draw as a two-dimensional image has no meaning and the constituent stars have no relationship with each other since if viewed from somewhere else in space the shape of the constellation would be entirely different.

The 12 regions of the Zodiac, as mentioned above, together make up a period of one year, and each single region therefore consists of a period of about 30 days. As already discussed, astrologers use the positions and movements of the planets within the constellations as a way of predicting personal events and personal characteristics. However, as will be apparent from the 2D and 3D diagram, saying that the planets are in a constellation conveys a completely inaccurate image. The implication is that 'in the constellation' means the same as, for example, your hand is 'in your pocket'. This is completely wrong.

The furthest planet, Neptune, is about four light hours from Earth, whereas the stars in the constellations are typically hundreds or thousands of light years from Earth. Although it would have appeared to the ancients that the planets were in the constellations due to the two-dimensional view, we now know that the planets have no connection whatsoever with the stars in the constellations. They are merely passing in front of them.

Here is a statement from an astrological website:

"The Moon spends much of its time in Taurus and then enters Gemini at 7.07 pm."

As we have seen, this is complete nonsense; the Moon isn't in Taurus any more than your finger is in the top of a distant building if you are pointing at it.

Reading horoscopes for fun is one thing but actually believing them is quite another. How could they work? The patterns and juxtapositions are irrelevant, and the influence of far-away planets and even more distant stars is hard to understand. Moreover, there have been several scientific papers published in reputable journals where properly controlled double-blind trials have shown that astrology does not work. For example, see "A double-blind test of astrology," *Nature*, 5 December 1985, 318: 419–52.

> **THIS IS AN ANSWERABLE QUESTION**
>
> **Answer**
>
> Astrology is based on several false premises and offers no plausible mechanism for its results, which are in any case often vague and ambiguous, and would apply to many people. Our answer is that it has no scientific basis, produces meaningless results and should be used for entertainment purposes only. Reports that some senior public figures consult astrologers for advice are deeply concerning, and hopefully untrue.

Q13

WILL WE EVER KNOW EVERYTHING?

Imagine you are a fish swimming about near a beach. You have eyes so maybe you perceive some sort of image of this dry stretch of land that has lots of creatures lying down on it. But what you certainly are not aware of are the digital cameras and computer tablets and mobile phones in these creatures' hands. Although these items may only be 400 metres away, they are separated from you by more than a mere 400 metres in distance; they are also separated by 400 million years of evolution.

Image 142. *Possible fishes' view of a beach.* ©*Author.*

So, are there things in the universe that are separated from Man in a similar way? Are there things out there that we do not have the capacity to understand? And are there things still to be invented that we cannot even dream about now?

These are hard questions to answer either scientifically or philosophically. We certainly know a great deal more than we did even ten years ago, and the sum of our knowledge is increasing ever more rapidly due to better and more powerful equipment.

It is probably safe to say that, in certain limited areas, we may well eventually know everything. For example, the field of particle physics is well-funded and is a busy area of research. It is entirely possible that in perhaps a hundred years or so we will know everything about the fundamental particles of our universe. Similarly, with medicine and genetics, both of which are also well-funded and well-researched, it is a fair guess that in a similar timescale we will know everything about all diseases and how to treat them.

These however are just linear projections from current knowledge. Some things are predictable in this way even though their implementation may require new technology. For example, an ancient Egyptian fisherman living 5,000 years ago might have watched birds flying gracefully overhead as he and his colleagues were at work, and pondered whether Man might ever be able to fly. His idea was a linear progression from what he saw even though it took until the early 19th century for Sir George Cayley to develop a glider that could carry a human passenger.

It is easy enough to predict developments of existing technologies, and the step from bird flight to manned flight is really in this category. But what are hard to predict are new concepts.

Image 143. *Ancient Egyptian fishermen and some birds (modified from an illustration on www.merchantnetwork.com.au).*

Our ancient Egyptian would not have been wondering about mobile telephones, digital cameras, radio and television, computers and the internet because these were not predictable from contemporary knowledge and experience.

In the same way, the future may bring us entirely new concepts that we cannot imagine now. And of course, when and if that future comes, the same could then be said about that future's future. So, on that basis, no, we will never know everything.

Finally, just like the fish swimming near the beach do not have the capacity to understand, say, a mobile telephone, is it possible that we do not have the capacity to understand certain things that may be around us? It would be rather arrogant of us to deny this, since some aspects of the design and function of the human body could be improved quite significantly.

Here are just a few ideas.

EYES THAT ARE SENSITIVE TO THE ENTIRE ELECTROMAGNETIC SPECTRUM

The electromagnetic spectrum is the entire range of wavelengths from long radio waves (from about 1 metre up to 1 kilometre) down to short gamma rays (in the range of

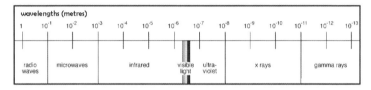

Image 144. The electromagnetic spectrum.

a million-millionths of a metre). Although there is a little variation in the eyes of different creatures, generally eyes are sensitive only to a relatively narrow range of wavelengths known, obviously, as the visible spectrum, as indicated in the drawing above. Seeing the entire spectrum like this makes it clear how much we cannot see.

But what if our eyes could perceive other wavelengths? What else could we see? Inside solid objects like Superman with his X-ray vision perhaps?

Superman was always drawn beaming X-rays onto his subjects as if he had an X-ray generator in his head. However, this is not how X-ray machines work. The X-ray image is captured on film or other sensitive material after the rays have passed through the subject. It makes much more sense to imagine that Superman had a retina that was sensitive to X-rays. If that were the case, then any background radiation of X-rays that fell onto his subject would be deflected according to the solidity of the subject, and, like an airport luggage scanner or medical X-ray, his eyes, if they were

Image 145. Superman and his X-ray vision. ©DC Comics with permission.

sensitive enough, could then detect items hidden from view to ordinary humans. (There are quite a lot of articles on the internet as to how Superman's X-ray vision could work). There are also audio frequencies that humans cannot hear but several animals can, so there are sounds out there of which we are unaware.

What about radio waves? We can get some idea of what seeing radio waves could mean by comparing images from optical and radio telescopes (Images 146 and 147).

Look at all the extra detail in the radio image. (Special image processing enables the radio data to be converted into a visible image).

It is clear therefore that we are only seeing a very small part of the world as our eyes are only sensitive to quite a narrow range of wavelengths. We have become used to

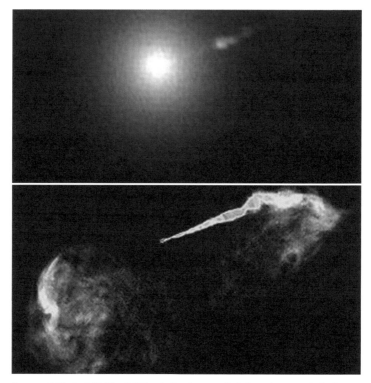

Images 146 & 147. *The M87 nebula photographed through an optical telescope (top) and by a radio telescope (bottom). ©NASA (top) and National Radio Astronomy Observatory/ National Science Foundation (bottom).*

dealing with this, and it serves our purposes, but how much better might it be if we could 'see' all the other wavelengths?

EYES WITH ZOOM LENSES

Anyone familiar with modern cameras will know that zoom lenses act like telescopes and bring distant objects closer. They achieve this by altering the focal length of the lens, and this requires a complex arrangement of separate lenses that move relative to one another during the zooming process. No living creature has this capability.

Images 148 & 149. *Aiming at a target with normal vision, and with telescopic vision. ©Author.*

Imagine trying to hit a target 50 feet away with a bow and arrow and then imagine how much easier it would be if you could zoom in on the target. Or, if you were an animal, how much more useful it would be to be able to spot prey.

PHOTOGRAPHIC MEMORY

The ability to recall information in extreme detail after a brief exposure is usually referred to as photographic memory; the technical name is eidetic memory. Examples would include the ability to recall the exact design of a large crossword or the names and positions of dozens of objects on a tray. This ability should be distinguished from feats such as learning the order of a shuffled deck of cards, since these almost always rely on a system of mnemonics.

Provided that one could choose when to bring this memory into play, it would clearly be an extremely useful skill that is currently enjoyed by only very few people.

THE ABILITY TO BIOSYNTHESISE VITAMINS

Vitamins are substances that are needed in small quantities by organisms to grow and remain healthy. Different organisms have different requirements; humans need 13 vitamins (A, B1, B2, B3, B5, B6, B7, B9, B12, C, D, E and K). Implicit in this definition is the fact that the human body cannot make these substances and therefore has to obtain them from its diet.

A sufficient supply of vitamins is essential for good health, and a shortage can lead to illness and death. Well-known examples are vitamin C (ascorbic acid) and its deficiency disease scurvy and vitamin D (calciferol) and its deficiency disease rickets.

Interestingly, guinea pigs are like humans in this respect since unlike most other small animals, they also cannot produce their own vitamin C so need to have it added to their diets.

Although most people with an adequate diet are very unlikely to become deficient in any vitamins, this is far from the case in poorer nations where deficiency diseases are rife. But if the human body could make (biosynthesise—produce by the body) all the vitamins it needs, then there would be no deficiency diseases.

Somewhere along the evolutionary path we have lost our ability to produce these substances ourselves.

IMMUNITY TO INFECTIOUS DISEASES

Apart from war, infectious diseases are the most significant cause of death throughout the world, with the poorer nations suffering the most due to lack of adequate hygiene and medication. For example, the influenza pandemic of 1918 killed between 25 and 50 million people worldwide and even today influenza kills about half a million people each year.

Animals and humans can be given immunity to certain infectious diseases by vaccination (from the Latin *vacca*—cow, since the first vaccine was derived from a cow).

This is a very effective process, so much so that the deadly disease of smallpox was officially declared as eradicated on 9 December 1979 and endorsed by the World Health Assembly on 8 May 1980.

If humans and animals were all immune to deadly infectious diseases as a matter of design rather than as a result of inoculation programs, then many millions of lives would be saved.

Obviously, there are many other improvements that one could think of, but the point is to show that at our present stage of development, humans are not perfect creatures in every way. If these imperfections extend to our brain power, then maybe there are things that, like the fish, we just cannot understand.

So, will we ever know everything? Even if one day we think that we do, how will we know that it is everything?

THIS IS AN UNANSWERABLE QUESTION

Best Guess Answer

Although we can have some confidence in what might happen in the near future, it would be supremely arrogant to imagine that we could predict events into the far distant future. Some of these events could even be beyond our current understanding, making it impossible to conceive what they might be.

It seems unlikely that we will ever know everything, but in a few million or billion years, who knows? As of now, we really cannot say.

Q14

THE CREATOR QUESTION

This is probably the biggest mystery of all and has deliberately been left till the end. In a way, it is a different type of question from the others in this book since one cannot discuss evidence in favour of or against as one can with all the other questions and that is why the chapter title has not been formulated as a question. Of course, some may say that the existence of the universe, and of life within it, is all the evidence you need to prove that a Creator exists since how else could it all have come about? But that is a religious view and this is a science book.

In Question 1, I concluded that since the universe had an origin and has not just always existed (not proven of course), invoking a Creator does not help us understand the mechanism of its origin and just gives it a name. That offers no help, so why do it?

In Question 2, we discussed what is sometimes known as the fine-tuned universe. This is a reference to the fact that the values of the physical constants, or if you prefer, the laws of physics and chemistry, are just right for galaxies, stars, planets and organic life to exist. Computer modelling has shown that if some of the values were only slightly different, a stable universe could not exist (just like the train in Question 2 that would run off the rails and crash if the track gauge was less than 1 percent wider or narrower). One might then reasonably enquire how the constants have the 'right' values. Many would say that they were designed that way by a Creator.

There are however other interpretations. There could be vast numbers of other universes with different laws, and we are in the one with the 'right' laws.

Consider five flower pots, each filled with different ingredients (Image 150). Only pot 4, with soil, has a flower growing in it since the others have the wrong ingredients to sustain growth. The flower, if it could think, might consider itself lucky to be in pot 4 but luck has nothing to do with it. Pot 4 is the only pot it could be in since the others are incompatible with life.

Or, one could argue that the laws could in fact be different but that a change in one is immediately compensated for by changes in the others. Going back to the train analogy, if the track gauge widens, and then the wheel distance also widens in a compensatory way, there would be no crash.

In Question 3 we considered the origin of life on Earth. How this happened is not known although plausible scenarios

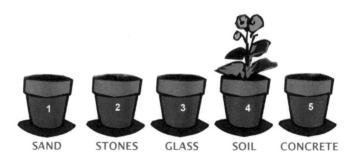

Image 150. Flower pots with different ingredients.

can be suggested, based on various scientific experiments. Again, just because something is unknown does not mean that there is no explanation that obeys the laws of nature. Invoking a Creator does not help us to understand the process—to repeat, it just gives it a name.

ORIGIN OF THE CREATOR BELIEF

Let us consider how the Creator or God option might have arisen. Imagine for a moment that you are a Neanderthal Man walking around somewhere in what is now present-day Europe.

It is 100,000 BCE and you do not know much about the world around you. You know how to make basic tools, how to kill and eat animals, how to seek shelter in a cave, and you know that it gets light and dark, and hot and cold.

You have also noticed a big and bright object in the sky that seems to appear, move across and then disappear. You have connected its presence with warmth and light and its disappearance with cold and darkness, and have become quite fascinated by it.

Image 151. Artist's impression of Neanderthal Man. ©Hunta-Fotolia.com.

Then one day, while gazing up at it, something simultaneously amazing and terrifying happens. It starts to disappear when it should not. It is like something is taking bites out of it. It starts to get very cold and very dark. You are afraid. Your heart beats faster and your breathing gets deeper. You start to sweat and panic. Then, after a few minutes, everything goes back to normal again.

We know this as an eclipse of the Sun. All that has happened is that the Moon has passed between the Earth and the Sun, temporarily blocking out its light and heat and obscuring it from view.

There were plenty of other things to frighten and worry our ancestor—thunderstorms with bolts of lightning, erupting volcanoes, tidal waves, hurricanes, etc. What would prehistoric Man have made of these events? Maybe he would have thought that they were a punishment for a bad deed or a warning.

These events would have been experienced time and again during Man's evolution, and at the time of the first civilisation around 6,000 BCE, usually ascribed to Mesopotamia in present-day northern Iraq, they were considered to have some purpose.

The Mesopotamians and other ancient civilisations tended to worship many gods, individually responsible for separate things such as love, death, the Sun, the Moon, the weather, etc., and were created as physical entities and/or paintings.

It is easy to see how populations might have derived peace and comfort from such worship when confronted with a major and inexplicable disruption to their normal day. The transient nature of astronomical and climatic events meant that things eventually returned to normal. For them, this was a persuasive indicator of the power of their prayers. Of course, had they done the control experiment, where there were no prayers the next time, the future of the God concept might have taken a different turn.

The concept of just one super-God responsible for everything probably arose during the later parts of the ancient Egyptian civilisation leading eventually to the widespread development of monotheistic religions.

It is probably a reasonable assumption to say that the God concept arose as a security blanket for ancient civilisations when confronted with unknown and worrying situations.

Those who choose to believe in a Creator will credit him/her/it with two tasks:
1   The creation of the universe and everything in it
2   Listening to and answering prayers

It is certainly difficult to explain how the universe could have been formed out of nothing, assuming of course that it had not been there forever. However, as was pointed out in Question 1, and illustrated with my card trick, just because we do not understand how something happened does not mean that there is no logical explanation that obeys the laws of nature. It merely means that no one has worked it out yet. Saying that it was done by a Creator merely gives the process a name but does not help us understand it.

Question 11 deals with prayer and considers whether it has any effect. Those who pray believe that their thoughts and words are somehow transmitted to the putative Creator who, they hope, will act upon them.

It is easy to find those who report positive outcomes after prayer (an ill relative recovers; a trapped explorer is rescued; a long journey is completed safely; perhaps even a major lottery is won). But that is not science—it is just cherry-picking positive results and is meaningless. Proper science would need control populations as described in Question 11.

Several such experiments have been carried out and published in reputable scientific journals, and have found no evidence that prayer has any effect on outcomes apart from, in some instances, a placebo effect.

SOME DIFFICULT QUESTIONS

To say that something has been designed rather than created naturally implies that some special expertise has been employed during construction to ensure that the article fulfils its function in the best way possible, given contemporary technology and resources.

If the universe and everything in it was indeed designed by a perfect Creator, then we can ask some questions about the details of the design. Here are a few suggestions:

- Why make the speed of light unsurpassable and the universe so big that travelling interstellar and intergalactic distances becomes impossible?
- Why design stars, including our Sun, so that they will eventually run out of fuel and all life in their solar systems will die?
- Why have asteroids and meteorites and comets which can, and have, destroyed many species of life with mass extinction events due to collisions?
- Why design a genetic process that does not have a fail-safe and can subject its carriers to the misery of incurable genetic diseases?
- Why design humans with personalities that prevent all people from living in harmony?
- Why allow millions of people to die in wars?

The classic answer to such questions is that we cannot and must not question God's motives. But that is not an answer; it is a cop out because there is no answer, except of course to conclude that God, as an all-powerful, all-loving and all-knowing entity, does not exist.

MY PERSONAL CONCLUSIONS

The creation of the universe and the creation of life are just processes obeying the laws of physics and chemistry. We may not fully understand them but, like the card trick described in Question 1, that does not mean that there is no proper explanation. We just have not worked it out yet. So, were the universe and life the work of a Creator God? Only if you wish to use that name for the processes by which these events happened.

But they are just processes. There is no justification for elevating them to a supernatural status and imbuing them with absolute power over everything that has happened and will happen. And there is also no justification for assuming that they will be able to receive and respond to prayer.

I believe that in due course the universe and everything in it will be completely explainable by contemporary scientific principles.

APPENDIX

A. EVIDENCE AND HOW TO ASSESS IT

The chapters in this book each consider a specific question and end either with an answer or, if no answer is available, a best guess answer based on current knowledge. This knowledge comes from scientific evidence, so it is worth discussing briefly the nature of evidence and when we should believe it and when we should not.

Evidence is information used to determine whether something is true or not. It is an essential component in all walks of life. Let us consider a few examples.

You pay for something in a shop and receive your change. You complain that the change is incorrect since you gave the shopkeeper a £50 note which you recall had a corner torn off. The shopkeeper insists you gave him a £10 note. On examination, the cash tray is found to contain a £50 note just as you described. The shopkeeper says that the cash tray had previously contained another £50 note with a corner torn off but although this is theoretically possible, most people would discount this as extremely unlikely. By far the most reasonable conclusion therefore is that you were right and the shopkeeper was wrong.

If you look at the sky and observe the motion of the Sun during the course of a day, you will notice that it rises in the

Image 152. Extremely unlikely. ©Author.

east, travels across the sky and sets in the west. The obvious conclusion from this observation is that the Sun moves around the Earth, resulting in Ptolemy's second-century geocentric model of the Solar System with the Sun (and planets) all revolving around a stationary Earth.

Although there had been earlier suggestions of a heliocentric Solar System, it was not until 1543 that Copernicus published his work showing that the Sun was at the centre of the Solar System with all the planets moving around it. Such an arrangement was the only one which complied with all the observations not only of the Sun but of the planets as well.

So here we can see that an apparently obvious conclusion can turn out to be false once additional evidence (the motions of the planets) is considered.

You have decided to visit a clairvoyant on the recommendation of a friend. You have recently lost your mother and you would like to contact her. After the visit your friend asks you how you got on.

"She was fantastic. She knew that my mother had died and told me that she still loved me and thought about me, and told me to look after the kids and never forget her."

Perhaps you should not be so easily impressed. You would have looked a little sad as you were thinking about your recently-deceased mother, so a guess at a lost parent, perhaps after one or two preliminary questions, is not so remarkable.

Based on your own apparent age, the clairvoyant took a shot that you had already lost your father—they tend to die first—and so guessed it was your mother. The rest of the 'reading' is pure generality and would apply to practically everyone in your position.

There are two ways in which the clairvoyant could have gathered the knowledge required to present your reading (although very little actual 'knowledge' was needed). Either she really did have psychic powers enabling her to make contact with dead people or she combined a couple of educated guesses with some generalities.

There is a name for this process. It is called cold reading. Skilfully performed, it can achieve seemingly remarkable results. Be reasonable; which one sounds more likely?

The first example shows that it is still possible to mount a defence against seemingly irrefutable evidence, but the stronger the original evidence the more fanciful the defence needs to be.

Who would believe that the shopkeeper already had a £50 note with a corner torn off in his cash desk? It is possible of course but so unlikely that very few people would believe him.

The second example about the Sun and the Solar System shows that incomplete evidence can lead to false conclusions which in this case had persisted for well over 1,000 years.

The third example of the clairvoyant shows that, faced with two explanations, many people will go for the most outrageous if that is the one they want to believe. Their desire to believe overrides their common sense.

It is like the story of the Cottingley fairies. These were a series of photographs taken between 1917 and 1920 by two young schoolgirls who lived in Cottingley in Yorkshire.

Frances Griffiths (10) and her cousin Elsie Wright (16) took the photographs as a joke.

They cut some pictures of fairies out of a book, propped them up in their garden with hatpins, and posed with them for a series of photographs, one of which is shown below.

Although never intended as a deception, the photographs became public and sparked a great deal of interest. Sir Arthur Conan Doyle, who was a spiritualist and believed in psychic phenomena, was convinced that the photographs were

Image 153. The first of several 'fairy' photographs, 1917. This shows Frances Griffiths. ©National Media Museum/SSPL with permission.

genuine as did other prominent people of the time. It was not until many years later, in 1983, that Griffiths and Wright admitted that the photographs had been faked.

It seems bizarre that two children could create a series of rather crude fake photographs and fool such eminent people as Sir Arthur Conan Doyle. The reason was that Conan Doyle and the others wanted to believe. This desire was so strong that it became the overriding factor in their decision.

It is important therefore to consider evidence on its own merits and not be swayed by what you would want the answer to be. Evidence forms an essential component of law and of science. It rarely provides an absolute truth since there is usually an alternative explanation and it is up to the observer to balance the probabilities and come to a conclusion.

Imagine that you are a member of a jury. You are shown a photograph of the suspect taken by a CCTV camera outside a house that has been burgled. Of course, many innocent people would have walked past the same location but you take one look at the photograph and immediately make a presumption of guilt. Why?

Image 154. *Villain?* ©*Author*.

Because he looks like a villain: dour expression, unshaven, hooded. But that is not evidence—that is prejudice.

Evidence would be his fingerprints on the stolen items or shards of house glass on his coat or his DNA in blood found at the scene. That is how verdicts are arrived at in courts of law and the same procedure should be adopted here when we come to consider the questions posed in this book. Preconceived ideas and wishful thinking have no place; it is the evidence that counts. And if there is no evidence? Innocent until proved guilty; if there is no evidence for X then all we can say is exactly that—there is no evidence for X. However, we should also remember that absence of evidence is not evidence of absence. In other words, just because there is no evidence of, for example, alien life, it does not mean that there is no alien life. It just means that we have no evidence of it.

We should mention the concept of Ockham's Razor. William of Ockham (also sometimes spelt as Occam) was a 14th-century English friar and philosopher who developed the concept that the explanation of any event should make as few assumptions as possible. In other words, the simplest explanation is likely to be the best one.

It will be obvious that some questions are easier to answer than others. This has nothing to do with a particular question being difficult but rather with whether the information to answer it is available.

SUMMARY

When deciding whether something is true or not, one considers the evidence. Sometimes it is clear-cut and there is little if any doubt about the answer, and sometimes one has to make a choice based on one's interpretation of the evidence and the balance of probabilities. Even though it may be difficult to do, one should not let bias and preconceived ideas get in the way of an impartial consideration of the available facts.

B. THE COSMIC DISTANCE LADDER— HOW ASTRONOMERS MEASURE DISTANCE

Knowing the distances to planets, stars, and galaxies is crucial to the understanding of the universe and the way in which these secrets have been unlocked is one of the great stories of science and is well worth reading about. There are many articles on the internet; there are also books— *Universal: A Journey Through the Cosmos* by Brian Cox and Jeff Forshaw (2017) gives a good description together with some interesting history.

Entire books can and have been written on this topic, much of which involves some complex mathematics and physics. This chapter is only intended as a brief outline of the subject in an attempt to explain the basic principles. Some basic mathematics is unavoidable though.

As will be seen, different methods need to be applied as the distances get larger and this has become known as the cosmic distance ladder, with each rung representing another measurement principle to go further than the previous one. It is important that the rungs overlap and that they start with an absolute measurement since this then enables each successive step to be calibrated by the previous one.

First Rung—Triangulation and Parallax

It is easy enough to measure distances in everyday life. Short distances can be determined with rulers and tape measures;

longer ones by looking at the mileage on a car odometer for example.

Things are not quite so easy when we come to objects that we cannot get to, either because they are inaccessible, such as a ship out at sea, or because they are very far away, such as the stars. However, simple geometry and some basic measurements can help us here.

The basis of the method is that if the three angles of a right-angled triangle and the length of one side are known, the lengths of the other two sides can be calculated. This is known as triangulation and is described in the example that follows.

For this method to work, we need to create a right-angled triangle. Let us imagine that we want to measure the distance to a tree from a nearby fence (the distance BC in Image 155) but that there is a shark-infested river between them that prevents us actually going to the tree. An observer at point A measures the angle α with a surveyor's theodolite. This is basically an instrument with a telescope and a protractor to measure angles. The telescope is aimed at point B parallel to the fence, and then swung round to aim at the tree. This then gives the angle α. Let's say it is 40°.

The distance AB, usually called the baseline, is easily measured with a tape measure or yardstick, and the angle ABC is a right angle because that is how it was drawn (and in any case, the perpendicular distance from the fence to the tree, the line BC, is the measure that we want). If we also assume that the distance from A to B was measured as 10 metres, we have all the information we need to work out the distance from the fence to the tree, the line BC in the diagram.

Trigonometry, from the Greek meaning 'to measure

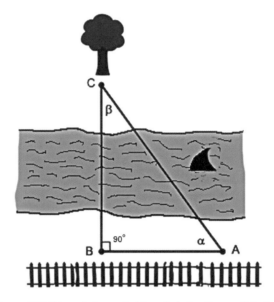

Image 155. Triangulation method for calculating distance. ©Author.

a triangle' was developed by Greek mathematicians and astronomers in the third century BCE. They noticed that the lengths of the sides of right-angled triangles and the angles between them had fixed relationships. This meant that if at least the length of one side and the value of one angle were known, then all the other angles and sides could be calculated.

These relationships between sides and angles were developed into reference tables as well as the definition of ratios, the most common being sine, cosine and tangent.

Trigonometry tells us that the tangent of angle α is equal to the length of the opposite side (BC) divided by the length of the adjacent side (AB) in a right-angled triangle. This would be written out as an equation:

$\tan \alpha = BC / AB$

Trigonometric tables, or the internet, tell us that the value of tan 40° = 0.84. Putting the values we know into the equation gives us:

0.84 = BC/10   so that BC = 8.4 metres

In the same way, surveyors can measure the heights of buildings or mountains without having to climb to the top. An angle and a length, and a right-angled triangle, are all that are required.

Parallax is a modified form of triangulation. The basic principle is the same with one addition that makes it a better method for astronomical use. This is because planets, and especially stars, are so immensely far away.

Imagine the tree and fence diagram but with the tree now 100 million miles from the fence. The length of BC would be 15 billion times longer than the 10-metre length of AB, and the triangle would be very stretched out. Angle α would be very close to 90°, so close in fact that it would require quite exceptional precision to measure it accurately enough. This of course means that the angle β would be exceedingly small, so small (in the micro arcsecond range—see notes on angles below) that it would also require exceptional precision for an accurate measurement. And although 100 million miles sounds a long way it is practically next door in astronomical terms.

So that is not going to work. A way around the problem would be to increase the length of the baseline (AB) which would then open up the angle β. But the increase in the baseline would need to be substantial to make the measurements possible. Perhaps the longest baseline we can imagine would be the length of the Earth's orbit around the Sun, so that measurements would be taken at 6-month intervals. This is shown in Image 156.

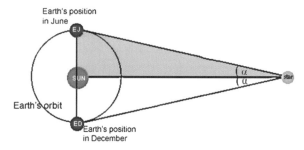

Image 156. *Diagrammatic representation of the Earth's orbit around the Sun and how to measure the distance to a star. EJ is the Earth's position in June and ED is its position in December. The line between EJ and ED is twice the distance from the Earth to the Sun. Modified from Wikipedia.*

Careful measurements made 6 months apart will enable astronomers to measure the angle α. Stars are so far away that they can be considered as fixed against the background. All that is needed now to determine the distance to the star, the line from the Sun to the star, is the distance from EJ (or ED) to the Sun. Trigonometry will provide the lengths of the sides of the shaded triangle as long as we know at least one of them.

That however immediately presents another problem – what is the distance from the Earth to the Sun (the baseline of the shaded triangle EJ to SUN)?

In the early part of the 17th century, after numerous careful observations of planetary motion, the German astronomer Johannes Kepler published his laws of planetary motion. His third law states that the period P (the time taken by a planet to orbit the Sun) is proportional to its distance D from the Sun. Specifically, the period squared is proportional to the distance cubed, or in equation form:

P^2/D^3 = constant

This of course implies that the ratio P^2/D^3 is the same for all the planets, Earth included. For Earth, the length of the period (year) is 1, and if we call Earth's distance from the Sun 1 as well, then the above ratio for Earth is also 1.

The important consequence of this is that we then know that the ratio will also be 1 for all the other planets meaning that we can work out the distance from the Sun of all the planets in terms of the Earth's distance as long as we know how long each one takes to orbit the Sun. Fortunately, this information has been known since ancient times.

So, for example, it was known that Mercury took 88 days to orbit the Sun, which is 0.24 of a year. Using the above equation, we can write that

$(0.24)^2 / D^3 = 1$

Working this out we see that $D = 0.39$ meaning that Mercury's distance from the Sun is 39 percent of Earth's distance. However, we still need to work out the absolute distances, in miles or kilometres, and for that we need to know the absolute distance of the Earth from the Sun.

It turns out that the Earth–Sun distance is a fundamental quantity that we need to know in order to determine all the other distances to the planets and many stars. It is so fundamental that it even has its own name—the Astronomical Unit abbreviated to AU.

There is a long and interesting history of attempts to measure the value of the AU, starting with Archimedes in about 250 BCE. It took almost 2,000 years until the first accurate measurement of the AU was made by the Italian astronomer Giovanni Cassini in 1673.

So how was this done?

We now need to introduce the concept of parallax. This is a well-known phenomenon, best demonstrated by holding a finger or other object close to your eyes and then closing

Image 157. Camera over right eye.   *Image 158. Camera over left eye.*

each eye in turn. The object will appear to move against a more distant background, as shown in Images 157 and 158.

The stick has not moved, only the camera, simulating looking with each eye separately. This apparent movement of a near object when viewed against a more distant background is known as parallax and it has been extensively used in astronomy to measure distances. The further away the background, the less the apparent movement of the near object.

Giovanni Cassini made the first accurate astronomical absolute distance measurement. This was the Earth–Mars distance, which he measured in 1673 by using parallax. Cassini realised that he needed to increase the baseline to get a larger parallax angle, so he sent a colleague astronomer to Cayenne in French Guiana while he stayed in Paris.

The distance between the two locations was about 4,500 miles (about 7,000 km). The set-up is shown in Image 159.

The two locations are marked A (Paris) and Cayenne, French Guiana (B). The lines of sight on the diagram show that each observer would see Mars (M) against a different

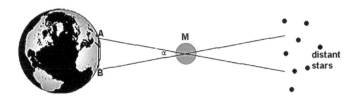

Image 159. Principle of Cassini's measurement of the Earth–Mars distance. Modified from Wikipedia.

background of distant stars. These are so much further away that they would appear fixed, enabling the angle of displacement of Mars (α) to be measured against them.

Drawing a line that bisects this angle and the baseline AB creates a right-angled triangle of which all the angles are known (the right angle, the measured parallax angle [½ α], and the angle MAB or MBA which are each 90 minus ½ α), plus the length of the baseline (half AB). This then enables the distance from Earth to Mars to be calculated by simple trigonometry as was done in the triangulation examples above.

Knowing the absolute distance from Earth to a planet (Mars in this case) then made it possible to calculate the distances to all the other planets by using Kepler's Law (pages 275–276) and, most importantly, the distance from the Earth to the Sun.

Here is the calculation that Cassini made based on Image 160 which shows the situation when his measurements were carried out. He chose a time when Mars was in opposition, that is, the Sun, Earth and Mars were in a straight line.

Cassini's parallax measurements yielded an Earth to Mars distance of 73 million km (about 46 million miles). This is y in the diagram. Kepler's third law tells us that Mars is 1.52 times further from the Sun than the Earth. In other words, $z = 1.52x$.

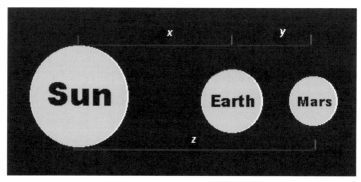

Image 160. *Diagrammatic representation of the Sun (S) / Earth (E) / Mars (M) positions for Cassini's measurements for the parallax of Mars. x = Earth to Sun distance; y = Earth to Mars distance; z = Mars to Sun distance. Not to scale. Modified from Wikipedia.*

Therefore, we have $z - x = 73,000,000 = y$, and we know that $z = 1.52x$. This gives us a pair of simultaneous equations which we can solve as follows:

$1.52x - x = 73,000,000$ which simplifies as
$0.52x = 73,000,000$

Therefore, $x = 140,000,000$

Cassini's careful work therefore established the distance from the Earth to the Sun as 140 million km (about 87 million miles), a result that is only about 7 percent away from the currently accepted values (150 million km; 93 million miles).

As we have seen, Kepler's third law of planetary motion now enables all the distances to the planets to be calculated, providing for the first time an accurate picture of the size of the Solar System.

Three hundred or so years later, the invention of the laser by Theodore Maiman in 1960 made possible a giant leap in the accuracy of these measurements.

By aiming a powerful laser beam at a moon or planet and measuring the time for the return journey of the pulse, one could compute the light travel time and hence the distance travelled by the light, which is of course twice the distance to the object.

The accuracy is astounding, of the order of millimetres. Unfortunately, this method is impractical for the stars, because even if a powerful enough beam was sent, the return time would be measured in years.

A note about angles: school trigonometry told us that a circle has 360° and most school protractors (Image 161), have sub-divisions down to 1°. However, each degree can be sub-divided into 60 arcminutes, and each of these can be further sub-divided into 60 arcseconds. There are therefore 3,600 arcseconds in 1°. Further sub-divisions down to milli- and micro-arcseconds are also used in astronomy.

Apart from the light year, the distance travelled by light in a vacuum in one year, astronomers also use another distance unit known as the 'parsec' which is equal to 3.26 light years. (Explanations of how this unit is defined can be found on Wikipedia or other science sources).

Now back to the stars. How far away are they?

The principle of parallax, and of observations made six months apart to create a large baseline, was used to make the

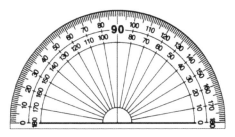

***Image 161**. Protractor.*

first accurate measurement of the distance to a star. This was achieved by the German astronomer Friedrich Bessel in 1838 who measured the parallax angle as 0.3136 arcseconds, which gave a distance to 61 Cygni as 10.4 light years (the modern value is 11.4 light years).

The parallax angle for 61 Cygni equates to about one ten-thousandth of a degree which gives some impression of the precision achieved by Bessel to make such measurements.

The Gaia spacecraft, launched by the European Space Agency in 2013 (actually from the same location, in French Guiana, where Cassini sent his colleague to measure the parallax of Mars in 1673), can make parallax measurements in the micro arcsecond range, resulting in distance determinations of up to 30,000 light years.

That is the current limit for making parallax measurements although one could perhaps imagine a future space mission leaving a parallax-measuring device on the surface of Pluto, thereby creating a baseline with Earth of up to nearly 3 billion miles (5 billion km). For the moment though, if we want to measure distances further than Gaia can achieve, we need to find another method.

SECOND RUNG—CEPHEID VARIABLES

Cepheid Variables are super giant stars that change their size and brightness in a regular way over a period of days or weeks, as shown in Images 162 and 163.

The first one was discovered in 1784 by the Dutch astronomer John Goodricke. In 1908, Henrietta Leavitt, an American astronomer, discovered that there was a relationship between the period (the number of days in the star's cycle of brightness changes) and its maximum brightness. The longer the period, the greater is the maximum brightness. For

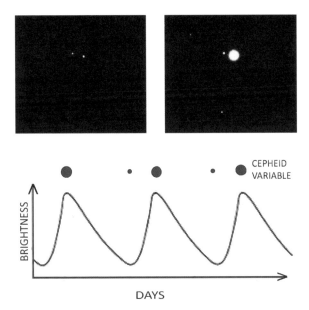

Images 162 & 163. Cepheid variable star at its smallest and dimmest and several days later at its largest and brightest (top); graph showing the regular changes in brightness (bottom). ©Wikipedia.

example, a Cepheid star with a period of say 20 days would achieve a greater maximum brightness than one with a period of only five days.

Leavitt observed these stars in a nearby galaxy which meant that, to a fair approximation, they were all the same distance from Earth. This meant that their difference in brightness was not due to a varying distance from Earth but was a true reflection of their actual brightness. The calculation itself involves some complex mathematics but the principle can be explained as follows.

It is obvious that two objects of the same brightness and at the same distance from an observer will appear equally bright, and if one is further away it will appear less bright. Light obeys the inverse-square law—its intensity decreases

Image 164. *Viewed from 10 metres. ©Author.*

Image 165. *Viewed from ? metres. ©Author.*

as the square of the distance. In other words, if a light source is twice as far away from an observer as a similar source, it will be ¼ as bright (inverse square, that is, 1 divided by 2^2, equals ¼).

For example, the car headlights shown in Image 164 were 10 metres from the camera and an unknown distance from the camera in Image 165.

Let us say that we use a light meter to measure the light intensity in both situations, and obtain the following results:

Image 164 = 50 lux   Image 165 = 12½ lux

So if we know the light intensity at a known distance (50 lux at 10 metres), and we know the light intensity at another but unknown distance (12½ lux at ? metres), the inverse square law enables us to calculate the unknown distance, which we can call D).

We can therefore write $50 \times 10^2 = 12½ \times D^2$
This becomes $5000 / 12½ = D^2$
so that $D^2 = 400$, and $D = 20$ metres

Various things are necessary for this method to work for the Cepheid variable stars as distance measures: a) We need to know the maximum brightness at a standard distance from the star so that we can compare this to the apparent maximum brightness of other similar stars at unknown distances; and b) we need to calibrate the method with parallax measurements of the same type of star so that we have at least one direct distance measure. The Danish astronomer Ejnar Hertzsprung made the first parallax measurement of a Cepheid variable in 1913 (delta Cephei) and found its distance to be about 900 light years. This then gave an absolute distance measure for a Cepheid variable and therefore calibrated the Leavitt formula which now enables absolute distances to be determined.

The Cepheid Variable stars are one of several light sources known as Standard Candles used by astronomers, meaning that they are of a known brightness at a defined distance (actually 10 parsecs—see notes earlier in this chapter) and can therefore be used to determine the distances of other astronomical objects in their vicinity.

Edwin Hubble famously photographed the Andromeda Nebula in 1923 (then thought to be a large conglomeration of

Image 166. Hubble's photograph of the Andromeda Nebula; note the barely visible star between the two straight lines (top right) above and to the left of the word VAR!. Hubble had first thought that this was a nova (an exploding star) and had written N next to it. He then realised, after more observations, that it was actually a Cepheid variable star and crossed out the N and wrote VAR! ©Carnegie Observatory.

ANOTHER UNIVERSE SEEN BY ASTRONOMER

Dr. Hubble Describes Mass of Celestial Bodies 700,000 Light Years Away.

Image 167. Press announcement of Hubble's discovery, 1925. ©New York Times.

gas within the Milky Way) and found a Cepheid variable star in one of Andromeda's spiral arms. A series of observations enabled him to observe the period of the star and then to calculate its distance as 700,000 light years.

Since the Milky Way was known to be about 100,000 light years in diameter, Andromeda must lie outside it and therefore be a galaxy in its own right. (Hubble's distance estimate was later found to be too small—Andromeda is actually about two and a half million light years away but Hubble had made his point. A little more on this topic is to be found in Question 1).

The Cepheid method can be used for distance measurement up to about 100 million light years but depends on being able to see the individual stars. Although this is possible for nearer galaxies, the vast majority are so far away that individual stars cannot be resolved. We therefore need another rung on our ladder.

Third Rung—Type 1a Supernovae

Supernovae are the end result of the explosion of a massive star as it ends its life. One of the most famous is the so-called Crab Nebula, observed and documented by Chinese astronomers in 1054. Although the remnants remain visible for hundreds or thousands of years, the initial light from the actual explosion fades away over several weeks or months.

A type 1a supernova occurs when one star in a binary system (two stars orbiting each other, one being a white dwarf) engulfs the other, gets too heavy and unstable, and explodes. Because the collapse of the star always happens when the same maximum mass is reached, the brightness of the explosion is always the same. Type 1a explosions are rare, estimated at about one per galaxy per 100 years. However, many have been observed in other galaxies and the distances of some in nearby galaxies have been measured by the Cepheid variable method.

Therefore, type 1a supernovae can be used as Standard Candles to estimate the distance to their host galaxies. They are so bright, up to several billion times as bright as the Sun, that they can be seen even though the star itself cannot be resolved.

Image 168. Supernova explosion. ©NASA.

Because the explosions are so very bright they can be seen and measured at distances up to billions of light years. One disadvantage is that the light from the explosion dims very quickly, often over a period of a few hours or days, so if a distance measurement is to be made, the explosion needs to be seen as or very soon after it happens or better still, to be seen even before it happens as a result of astronomical clues from previous observations.

Fourth Rung—Red Shift

Our final rung describes a method that has the potential to measure distances as far as we can see. It also provides proof that the universe is expanding. This method, known as Red Shift, depends on a property of the light reaching us from far-off galaxies. It is useful to have this method available since type 1a supernova explosions are rare events.

In 1671, Isaac Newton published his discovery that a glass prism could separate different colours out of white light, and used the word *spectrum*—Latin for 'appearance'—to describe the phenomenon. Then, in 1802, English chemist William Wollaston, noticed that the spectrum of the Sun contained a number of dark lines. Twelve years later, in 1814, German physicist Joseph von Fraunhofer rediscovered these lines and made a detailed study of their positions in the solar spectrum. After much study, he eventually realised that the lines were due to the lack of a particular wavelength of light in the spectrum, and that this was in turn caused by elements in the Sun that absorbed these particular wavelengths. Most significantly, each element had its own specific line or lines, so that the precise position of these lines could be used to identify the element.

This was a staggering discovery since it meant that one could identify the elements in the Sun (and of course any other light source) by examining its spectrum. For now, our interest lies in the so-called Fraunhofer lines and their precise position in the spectrum being examined. (This is how helium was discovered in the Sun before it was found on Earth; from Helios, the Ancient Greek god of the Sun).

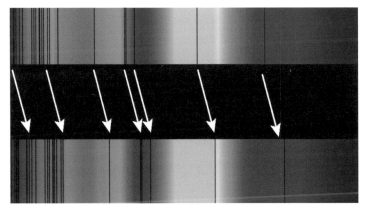

Image 169. *Solar spectrum (top) and distant galaxy spectrum (bottom) showing the positions of the Fraunhofer lines and how they have shifted.* ©*Wikimedia Commons.*

Image 169 is a photograph of the Solar spectrum (top). It is in black and white but what is important are the positions of the black lines. The lower spectrum is of a distant galaxy.

As is indicated by the arrows, it is clear that the positions of the lines have shifted to the right (the red end of the spectrum). This is called Red Shift. The actual pattern of lines remains the same; it is just their positions that have moved.

What has happened here is that the expansion of the universe has stretched the light waves and this is shown as a displacement of the absorption lines. Not surprisingly, there is much more physics and mathematics involved in turning these shifted absorption lines into distances but that is certainly beyond the scope of this chapter (and me) so we can jump straight to the results.

The Red Shift is denoted by the symbol z and it tells us the light travel time from the object being studied. This however is not the same as the distance to the object because the universe has been expanding as the light travelled, so the object is further away when the light finally reaches us.

So here is a table giving the light travel times and the proper distances now for different values of the Red Shift.

Table L. Red Shifts and distances

RED SHIFT	LIGHT TRAVEL TIME	DISTANCE NOW
1	8 billion years	10 billion light years
5	12 billion years	22 billion light years
10	13 billion years	27 billion light years

The most distant object so far discovered is a galaxy known as GN z11 which has a Red Shift of over 11, implying a proper distance of 32 billion light years.

While on the subject of distance in an expanding universe, it is worth mentioning that astronomers use more than one definition of 'distance'. We can elaborate on this a little as follows.

Since the universe is expanding, the distance from us to far away objects is increasing due to this expansion. It is a bit like continental drift whereby the distance from London to New York for example is increasing by about 2.5 cm (1 inch) a year due to the movements of the Eurasian and North American tectonic plates away from each other.

So, if the straight-line distance between these two cities is 5,585 km (3,470 miles) now, then it would be 5,585.000025 km (3,470 miles and 1 inch) in one year. It is easier however to ignore the extra distance due to the continental drift and to allow for it when necessary since its rate is known.

Similarly, in cosmology, scientists define several different distances when discussing far-away objects. The two commonest are the 'proper distance' and the 'comoving distance'.

The 'proper distance' is the distance now taking the expansion of the universe into account. In the London–New York example quoted above, it might be 5,585.01 km if the time since the original measurement was made has meant that the continental drift has added 0.01 km to the distance. Looked at another way, it is the distance that would be measured if one used a set of rulers to make the actual measurement between the two objects in question.

The 'comoving distance' ignores the expansion of the universe (or the movement of the tectonic plates) and, again in the above example, would be given as 5,585 km. Since the rate of expansion is known, this can be factored in afterwards if necessary. At time now, the proper and comoving distances are the same.

This chapter is intended as an introduction to the subject of distance measurements in astronomy. The examples given are meant to illustrate principles, since things are somewhat more complicated in reality. For completeness therefore, it seems appropriate to make the following comments.

The orbits of the planets around the Sun are not true circles but ellipses, and this needs to be taken into account when precise measurements are made of planetary distances and positions.

In practice, there are more rungs on the cosmic distance ladder than mentioned here. The really important point is that each lower rung can be used to calibrate the next one, and that the first rung provides actual distances. For more complete information, look up Cosmic Distance Ladder.

SUMMARY

The measurement of astronomical distances has a long history stretching back to at least the ancient Greek philosophers and scientists of the third century BCE. Eratosthenes, for example, was able to measure the circumference of the Earth without leaving his home town in Egypt.

Current information is obtained via the cosmic distance ladder, an overlapping series of measurements depending on different principles, each successive one reaching further but calibrated by the one before it. The first rung on the ladder, laser ranging and parallax, provide absolute distances and these form the basis for all the others. The most distant object seen to date is a galaxy about 32 billion light years away.

FINAL COMMENTS

I hope that you have found this to be a stimulating and interesting book even though you may disagree with some of my conclusions. There is nothing wrong with that since it can result in interesting discussions.

If there is one message that I would like to impart it is that people should not make decisions based on wishful thinking but should look at available evidence and use that to reach a decision, just like is done in courts of law. Do not just blindly accept fantastic ideas because you would like them to be true. Think about alternative explanations that could be much more likely.

Of course, there are many things for which there is no explanation at all, and several of these have been discussed in this book. That does not mean that there is no explanation—it just means that we have not worked it out yet (like my card trick example in Question 1).

I would be delighted to hear from any reader via the publisher about any of the topics in the book. Both complimentary and critical comments would be welcomed.

a science fiction short story in 3 parts

THE INESCAPABLE CONCLUSION

Peter Altman

THE INESCAPABLE CONCLUSION

PART ONE

It was 2.30 in the morning and the road leading up to Mission Control was quiet. Dr.Emil Petersen drove in as the barrier lifted and parked in his space. He was looking forward to his session tonight as he would be on duty when the first images came in. It would be the culmination of an 11-year project and tonight would make it all worthwhile. The lander had touched down safely on Ganymede, Jupiter's largest moon, and the sophisticated imaging package had deployed as planned. Emil switched on his monitor and sat back in his chair. Forty seven minutes to go.

Precisely on time the flickering monitor screen stabilised and Emil was privileged to be the first person to see a three-dimensional colour image of the moon's surface. The heavily cratered surface was immediately apparent but suddenly his eyes widened and he leant forward, hardly believing what he saw. But there could be no doubt. The object in the centre of the screen, with regular black and green stripes, was moving.

His heart was racing as he reached for the telephone to summon his colleagues but before he could speak the object had turned and begun moving towards the camera. Then the screen went black.

Twenty minutes later Emil had been joined by biologist Joy Gibson and astrophysicist Dan Greening. They had watched the 34-second video at least 20 times but no one had voiced what they were all thinking. Finally, it was Dan who said it. "Life. Alien life." "And those dark green and black stripes—perfect for absorbing the weak sunlight for some type of photosynthetic process."

The news made headlines worldwide the next day and Emil and his team were besieged by reporters. In a few hours, they were as well known as the Apollo 11 astronauts returning from the first Moon landing in 1969. Emil felt bathed in glory as the first person to see an alien life form, now affectionately known as 'Stripey'. Naturally, everyone wanted to see an image of the creature but Steve Morrell, the Mission Controller, had held them back because Joy, always a cautious worker, had begged him to do so. "Something's not right," she said. "I want to have another look."

"We can't hold them back much longer," said Steve. "You've got an hour."

While the rest of the team were being feted by the media, Joy rushed back to Mission Control on her own. She played the video again but this time in high resolution. Staring intently at the image, she spotted something that the others had missed.

They had been so intent on looking at the creature that they hadn't looked beyond it, or, to be more exact, below it. "Look at that," Joy thought to herself, "the thing isn't actually touching the ground. It's gliding over it elevated by some sort of gas pressure system."

Then a shiver came over her as the truth dawned. "That's not a living creature; it's a manufactured object. Someone's got to Ganymede before us."

An hour later the team were together in an urgent meeting with the Mission Controller. "This could be a complete public relations disaster," said Steve. "Once this gets out we'll be a laughing stock. It could even jeopardise our future funding. We should have been more cautious and never rushed out with this so quickly."

"Hang on a minute," said Dan. "If we didn't put Stripey on Ganymede then who did? We need to find out."

It didn't take long. Only very few countries have the capability to send space probes that far and it was soon confirmed that none of them were responsible. "There's an inescapable conclusion," said Steve. "Stripey may be a manufactured object but whoever built it and sent it isn't—and he doesn't live on Earth."

PART TWO

The Chairman stood up to address the entire staff who had crammed into the boardroom. Dr. Arnold Weiss had already decided on his course of action and this meeting was to direct and not to discuss.

"This is what we know," he said. "There is a device on the surface of Ganymede that was not put there by any nation on Earth. The inescapable conclusion is that it is of extra-terrestrial origin. We need to examine this device. Sending another mission is out of the question due to the time this would take. We will therefore reprogram the orbiting mother ship to make detailed observations of the device using every instrument and facility on board."

"From a purely scientific point of view we would want to know where it came from, how long it's been there, how it got there, what it's made of and what's inside it. However, our priority is to find out what it can do and if it's a threat. So that is what we will focus on first."

"We will have update meetings every two hours. And one

more thing; no one leaves the site and no one talks to the media. Now go."

Two hours later the Chairman called the first update meeting to order. "OK ladies and gentlemen, what have we got?"

Joy stood up. "Yes, Dr. Gibson, tell us," said the Chairman.

"I've been thinking about the stripes on the device," said Joy. "It would be pointless if they only had a decorative function so they must be there for a reason." The Chairman leant forward. "Go on."

"Well, Sir," continued Joy, "when we thought that the device was a living creature, Dan Greening suggested that the green and black colouration would be very useful for some sort of photosynthetic process but we abandoned the idea when we realised that it was a manufactured device."

Joy now had the attention of the entire room.

"But what if the device was dependent on photosynthesis for maintaining itself? What if it was a self-replicating device?"

Dan shot up. "That could mean that there are others."

Before anyone could comment the door burst open and Frank Peters, the Chief of the Imaging Team, ran in. "Sorry to interrupt, Dr. Weiss, but we've just received some staggering news."

"We've found another device. And it's on our Moon."

PART THREE

The discovery of a second device on the Moon and Joy's suggestion that Stripey might be a self-replicating robot was soon leaked to the media where it understandably caused widespread panic. Headlines such as 'Invaders from Mars' and 'War of the Worlds' appeared in every country. All radio and television channels, and all social media channels, carried the story. Even though official comments had been restricted to the basic facts and had deliberately excluded the idea of self-replicating robots, commentators had quickly created their own version that aliens were preparing to attack the Earth.

The original calm at Mission Control had been replaced by a frenzy of activity. The discovery of a second device on the Moon, while clearly creating great concern, did at least mean that a reconnaissance mission was feasible.

The Chairman wasted no time. "How long before we can get a team to the device?" he asked the Lunar Expedition leader. "Well," began Tom Crosby but he was interrupted by the Chairman before he could continue. "For all we know, our entire existence could be under threat. You've got four weeks. I don't care what it costs or what it takes, I want our people on the lunar surface in four weeks."

Three days later, Frank Peters, Chief of the Imaging Team, asked for a private meeting with the Chairman. "Dr Weiss, I've asked for this meeting with you, in the strictest privacy, because our team have made a truly devastating discovery and we don't know how to handle it."

"Go on," said the Chairman.

The Chairman fell silent for a moment after hearing what Frank Peters reported. Then he just said, "Thank you, Dr. Peters. Please return to your department."

"Ladies and Gentlemen. This is the Chairman. I have an important announcement. Please will you all stop what you are doing, whatever it is, and make your way to the boardroom immediately. No exceptions—everybody, now."

The Tannoy message reverberated around the building and had the desired effect. No one spoke as the Chairman rose.

"A discovery has been made which means that we now have to release the following statement to our Government. It is out of our hands."

He picked up his tablet and began to read.

"Four days ago, we discovered an alien device on the surface of Ganymede, Jupiter's largest moon. Yesterday we discovered a similar device on our Moon. Today, after computer analysis of archive photographs, we have discovered such devices on every planet's largest moon.

It is our inescapable conclusion that we are, or soon will be, under attack from an alien civilisation."

No one said a word.

"You will now all return home to your families. This facility is closed until further notice."

With that, the Chairman left the room.

The news spread like wildfire around the globe. The normal social order was replaced with sheer panic in the face of an unknown enemy with unknown powers. Amassing arms, escaping to higher ground, building shelters, hoarding food and water, etc., were all rightly considered valueless defences. Looting was rife, offering a final escape to a last-minute indulgence.

Long-standing conflicts based on religion, social class and disputed territories took second place to annihilation of the planet, and in less than a day the world, although socially bereft, was for the first time in recorded history united by a common threat and without war.

The last sentence can be found in the Index.

INDEX

abiogenesis zone 133, 134, 137, 138
Adamski, George 109–113
ageing 184–86, 189–90
age of the
 Earth 55, 57, 60, 69
 universe 10, 58, 207
alien encounters 81, 109. 118
alien life xi, 80, 90, 92, 120, 122, 148, 150, 156, 157, 159–61, 172, 270, 295
amino acids 39, 43, 45, 56, 126–27, 186
ancient astronauts 97, 101, 104
Andromeda 9, 58, 145, 153, 166–67, 174, 185, 284–85
anthropic principle 34
Archimedes 276
asteroid belt 194
asteroids 194, 197, 263
astrology 241
astronomical impacts 122, 194, 201–202, 218

astronomical unit 276
axolotl 189

Big Bang 11–12, 15–18, 76, 206, 213
Big Chill 207, 209–213, 219
Big Crunch 17, 206–209, 212–13, 219
Big Rip 206, 208, 212–13, 219
biomarkers 148, 150
Blue Origin 170
Borisov 95–96
Breakthrough Initiative 127–29

calorie restriction 186
carbon 32, 61–62, 72, 76, 78, 91, 137
card trick 26–29, 262, 264, 292
Casimir Effect 29
Cassini 276–78, 281
Cepheid variables 281–86
Chariots of the Gods? 96–97, 101
Chicxulub crater 198

circular aircraft 87
CMB (Cosmic Microwave Background) 12–13
COBE (Cosmic Background Explorer) 12
coincidences 223, 232–33
coin-tossing robot 139
comets 39, 194–97, 263
comoving distance 290
constancy of physical constants 35
constant acceleration drive 164, 171, 179
cosmic conspiracy 14
cosmic distance ladder 271, 291
cosmic inflation 13–15
Cottingley fairies 268
Creationism 53, 57
Creation 53–55, 57–59, 262–64
critical density 207–208
cryopreservation 188, 191

Dark Energy 23, 209–17
Dark Matter 23, 209–17
Darwin, Charles 48–49
date for creation 59
decay rates 70, 74–75
dendrochronology 78
different physical laws 36
distance measurement 277, 286–87
von Daniken, Erich 96–97, 112

DNA 43–48, 126, 270
Drake equation 123–29, 133, 139, 151
Drake, Frank 122–23

ecliptic 242–43
electromagnetic spectrum 12, 251
Elixir of Life 181, 186
EmDrive 164
end of the World 192
entropy 211
event horizon 8
evolution 50–56, 64, 157, 189, 248, 261
extraterrestrial 96, 106, 122, 129, 152
exomoons 155
exoplanets ii, 80, 90–91, 126, 131–32, 138, 148–50, 155
expansion of universe 6–8, 13, 79, 165–66, 206–209, 212–13, 216–17, 289–90

'Face on Mars' 102
faster than light travel 165–66, 170–71
Fermi paradox 157, 159
fine tuned universe 33, 258
Fish, Marjorie 115, 119
Fraunhofer lines 288
frequency of impacts 200
future of the universe 206, 218

302 Index

gamma ray burst 78. 203–204. 218
Goldilocks zone 47
Grandfather Paradox 175, 180
Griffiths, Frances 268

habitable zone 47, 90–91, 132–33, 137–38
half-life 60, 62–74, 78
Halley's comet 195-197
Hill, Betty and Barney 113–16, 118–19, 244
horoscopes 242, 246, 248
Hubble, Edwin 9–10, 35, 76, 129, 208, 285–86

ice cores 59–60, 70–73, 77
impact craters 198
inflation 13–15
inter-galactic travel 165, 170–72, 263
intelligent design 53, 57
in-transit creation 58, 77
isochron dating 67
isotopes 61–62, 68–78

Kepler spacecraft ii, 131–32, 138
Kepler's Law 214, 275, 278–79
Kuiper belt 161, 194–95

last sentence – *On its home planet, the alien creator of Stripey made a facial gesture that we would call a smile.*

Lazarus trial 188
Leavitt, Henrietta 281–84
lenticular clouds 88–89
life expectancies 181–82, 186
living longer 182–90
Lowell, Percival 120

M Theory 26, 28
Martian 'canals' 121
mass ratio, constancy 35
Melaris, Spyros 86
Mercury 20–23, 205, 213, 276
meteors 194
Milky Way 3–5, 9, 75–76, 125, 157, 285–86
Miller-Urey experiment 42–43
mitochondria 50
Moon 54, 74–75, 92, 94, 120, 130, 133–38, 149, 155, 160, 163, 169–71, 198, 203, 206, 229–31, 241, 246, 261
Moons and exomoons 155
multiverse 34, 36, 219

Neanderthal Man 260
Near Earth Objects 95, 199, 202
nuclear pulse propulsion 163

observable universe 6–8
Ockham's Razor 55, 77, 116, 229, 270
origin of life timeline 48–49
Orion Arm 4

Orion constellation 243–45
oscillating universe 17, 207–208
other universes 31, 34, 37, 259
Oumuamua 95–96

Panspermia 38
parallax 271, 274–84, 291
Parker Solar Probe 92
parsec 280, 285
Perfect bridge hand odds 143–44
phosphine on Venus 154
physical constants 30–37, 76, 258
physical evidence for aliens 92
Pioneer spacecraft 93, 152–53, 162
Pioneer plaque 93
Piri Reis map 103–106
placebo effect 235–37, 263
planetary systems 80, 145–46
progeria 184–85
proper distance 289–90

racemic mixture 45
radioactive decay and dating 59–62, 67, 74–75
rapamycin 187
red shift 287, 289
regeneration of organs 189
remote detection 148
re-vitalisation 188

right-handed and left-handed molecules 39, 45–46, 126–27
robot, coin-tossing 139
rock carvings 101
Roswell Incident 84–85

Second Law of Thermodynamics 209–10
senolytics 187–88
SETI 123–27, 145–46
solar spectrum 288–89
Space X 170
Spare part surgery 189
spark chamber 24–25
standard candles 285–86
string theory 26
Sun life cycle 151, 205, 218
Superman 251–52
supernova 75–76, 203, 286–87

telomeres 184–86
theory of everything 25–26
time dilation 176–80
time travel 173–80
Torino Scale 262
transit 131, 149
tree rings 60, 72, 77
triangulation 271–74, 278
trigonometry 272–73, 278–80
Tunguska event 200
type 1a supernova 286–87

UFOs 81–87, 109–115